Lucas Maciel de Aguiar
Sérgio Pacífico
Higino Matos

Capacity analysis at intersections: Case study in Ponte Nova-MG

Project to alter the circulation of vehicles in the central area of the municipality

ScienciaScripts

Imprint
Any brand names and product names mentioned in this book are subject to trademark, brand or patent protection and are trademarks or registered trademarks of their respective holders. The use of brand names, product names, common names, trade names, product descriptions etc. even without a particular marking in this work is in no way to be construed to mean that such names may be regarded as unrestricted in respect of trademark and brand protection legislation and could thus be used by anyone.

Cover image: www.ingimage.com

This book is a translation from the original published under ISBN 978-613-9-63017-2.

Publisher:
Sciencia Scripts
is a trademark of
Dodo Books Indian Ocean Ltd. and OmniScriptum S.R.L publishing group

120 High Road, East Finchley, London, N2 9ED, United Kingdom
Str. Armeneasca 28/1, office 1, Chisinau MD-2012, Republic of Moldova, Europe
Printed at: see last page
ISBN: 978-620-7-76461-7

Lucas Maciel de Aguiar
Sérgio Pacífico
Higino Matos

Capacity analysis at intersections: Case study in Ponte Nova-MG

SUMMARY

I dedicate this work to my parents, who have always believed in my potential and encouraged me throughout my life.

I thank God for his strength;

To my parents and my sister for their support and trust;

To my advisor, Sérgio Pacifico Soncim, for his attentive guidance, which helped me a lot in carrying out this work;

To the engineers Higino Matos and Heitor da Hora, for hosting me in Campinas and helping me with the *software*;

To Perplan Engineering, who supported me in obtaining the results;

To Bàrbara Marinho, who supported me and dedicated much of her time to helping me with the vehicle counts;

To the Ponte Nova - MG City Hall and the Municipal Traffic Department for their support in the field survey;

To my friends for their support.

SUMMARY

The objective of this work was to develop a traffic modeling proposal for the study of traffic signal alternatives that reduce congestion at the intersection of Av. Francisco Vieira Martins with Av. Dr. Otàvio Soares and its surroundings, in the municipality of Ponte Nova - MG. Because it is on the city's main avenue, vehicle traffic at this intersection tends to have a direct and indirect influence on the city's traffic volume. In addition, this work proposes a reflection on the reasons why this intersection receives a high volume of traffic and proposes a technically feasible solution to the problem in the short and medium term. To do this, it was necessary to evaluate the cycle of the traffic lights at the main intersection and its surroundings, find out the flow rates at the main intersection and its surroundings, and propose alternatives to reduce density and travel times. The work revealed the need for actions involving partnerships between managers and the local community, as everyone is responsible for the safety of individuals in their surroundings. It was understood that it is possible to optimize the use of the road network studied in this work, as well as enabling pedestrian traffic at intersections with greater safety.

Keywords: Traffic. Semaphore intersections.

1 INTRODUCTION

1.1 Initial considerations

Traffic is part of the daily life of the vast majority of people living in urban centers, and we can see that the time taken to complete the same journey has increased more and more compared to a few years ago. This problem is caused by the large number of vehicles on the streets, especially at peak times, where the tendency is to generate congestion, traffic accidents, environmental impacts, wasted fuel and, consequently, stress for drivers and passengers.

Capacity at intersections is directly linked to the supply of services, saturation flow and effective green and traffic light cycle times. Because the processes of urbanization and land use in developing countries are often accelerated and disorganized, they end up causing a greater demand for road space than the existing supply. This supply can be defined as the current road structure, which can often be used inappropriately as a result of poor planning, lack of organization and integration between the public sector and society.

The need to invest in a master plan that aims to rationalize urban space in the midst of reduced capacity and high demand is becoming increasingly evident. However, in order to implement a project of this magnitude, it is necessary to look at a number of variables that can alter land use, public transport and the entire political and ideological context.

All this can be solved with good planning. To do this, a prior capacity analysis is essential, not just of the intersections, but as a network in general. From there, solutions should be proposed that involve the construction of new road networks, in order to relieve congestion in the areas with the greatest flow, and also a reorganization of the existing road network.

As the construction of a new road network is often unfeasible due to costs and the physical and geological characteristics of the region's terrain, this project aims to redesign traffic at the intersection of Av. Francisco Vieira Martins and Av. Dr. Otàvio Soares, in the municipality of Ponte Nova - MG. This process can be carried out by reprogramming the existing traffic lights and changing the direction of the roads, with the aim of making the project functional and cost-effective for the relevant authorities.

1.2 Justification

The city of Ponte Nova - MG has been experiencing a great deal of traffic congestion, mainly at the intersection of Av. Francisco Vieira Martins and Av. Dr. Otàvio Soares. This study will be able to assess the level of service provided by the main intersection and propose improvements in order to optimize traffic in the municipality as a whole. The intersection has a huge influence due to the fact that it is in the Palmeiras neighbourhood, a place with a high concentration of commercial enterprises.

1.3 Objectives

1.3.1 General objective

The aim of this work was to develop a traffic modeling proposal for the study of traffic signal alternatives to reduce congestion at the intersection of Av. Francisco Vieira Martins with Av. Dr. Otàvio Soares and its surroundings, in the municipality of Ponte Nova - MG using *TransCAD, SIDRA and AimSun software*.

1.3.2 Specific objectives

- Evaluate the cycle of the traffic lights at and around the main intersection;
- Know the flow rates of the main intersection and its surroundings;
- Propose alternatives to reduce density and travel time;
- Reorganize the road network in order to obtain acceptable levels of service.

1.4 Organization of the Monograph

In addition to the Introduction, the monograph has three more chapters: Bibliographical review, which is designed to conceptualize the parameters necessary for understanding the project described in chapter three, as well as demonstrating the state of the art of the monograph, using works and articles related to the topic through citations; Method, consisting of a detailed description of the methods and procedures used in the development of the project; Results obtained, which refers to the conclusions, where the results will be demonstrated, with analysis, proposed solutions and suggestions for a possible continuation of the work; Conclusion and, finally, the references used to make the project and the annexes used.

2 LITERATURE REVIEW

2.1 Traffic Engineering

The science based on the movements of pedestrians and vehicles carrying people and goods is called traffic engineering. It studies movements, traffic lights, road hierarchization, organization and planning of the road system, analysis of the impact of major road works on the road system, as well as the relationship between public and private transport. The aim is to improve mobility within the road network analyzed, minimizing environmental and socio-economic impacts. Cucci Neto (2011) states that it is the responsibility of the traffic engineer to draw up signaling projects, aiming for safety, comfort and fluidity in travel.

The definition of traffic engineering described by the *ITE- Institute of Transportation Engineers* in the United States is:

> Traffic Engineering is the phase of Transport Engineering related to the planning, geometric design and traffic operations of roads, their terminal networks and adjacent land, including the integration of all modes and types of transport, with the aim of providing the safe, efficient and convenient movement of people and goods (SOARES, 1975 *apud* VASCONCELLOS, 1999).

2.1.1 History

In a brief history, the first manual traffic light made up of gas lamps was developed in 1868 in London. The first electric traffic light was created by James Hoge and applied in Cleveland later in 1914, giving rise to the three-color traffic light known today (HOMBURGUER *et al,* 1992 *apud* PIETRANTONIO; BONETTI JR, 2002).

However, in Brazil it is not known for sure when traffic lights began to be used. From the 1970s onwards, through the experience of other countries, the SEMCO traffic light control system was developed, of North American origin, consisting of a system of traffic light intersection controllers that act in fixed time from a central network, led by CET/SP - Controle de Engenharia de Tràfego de Sâo Paulo (HOMBURGUER *et al,* 1992 *apud* PIETRANTONIO; BONETTI JR, 2002).

2.2 Corrective measures

In *the* midst of chaotic traffic, measures are used to alleviate the most recurrent problems, such as: a) restricting access for large vehicles in certain areas of the city; b) improving public transport; c) increasing enforcement; d) improving urban planning; and e) restricting on-street parking or taking direct action in parking lots.

In addition to these measures, there are alternatives for improving signalized intersections such as: a) analysis of traffic light timings (they may have green and yellow times that are out of sync with the volume requested); b) a special phase for left turns; c) a ban on stopping and parking in the vicinity of intersections in order to maintain greater capacity; d) if possible, a reduction to a maximum of two phases per intersection.

2.3 Concepts

There are some technical concepts covered by the DENATRAN Traffic Light Manual, such as: a) movement: defines the origin and destination of the vehicle at the intersection. In order for the movements not to conflict, it is necessary to create different phases for each one b) cycle: time expressed in seconds of the sum of the green, red and yellow times; c) phase: sequence of colors applied to one or more movements during a cycle; d) stage: time in which the color of a certain traffic light does not vary and, e) time between greens: It is usually defined by the yellow time, but in some intersections it is used as the sum of the yellow time and the general red time (BRASIL, 1984).

According to the *Highway Capacity Manual - HCM*, which deals with road capacity around the world, some concepts should be considered: Traffic volume: the number of vehicles passing a fixed point at a given time. They are usually measured in hours (Hourly Volume - VH), days (Average Daily Volume - VDM, recorded in 24 hours), or years (Average Annual Daily Volume - VDMA, the volume of a day multiplied by 365); Peak Hour: Hour of highest traffic volume; Peak Volume: Number of vehicles corresponding to the peak hour; Peak Hour Factor (PHF): The peak volume divided by 4 times the highest consecutive traffic volume in 15 minutes within the peak hour. This factor is needed to adjust the intensity of the peak hour, which can be uniform or have very high values in some fractions of an hour and low values in others; Density: This is the number of vehicles on a given stretch at a given time. It is measured in vehicles/km; Capacity: It is the maximum flow that a given stretch of road can handle at a given time. The road should not be operated close to its capacity, as this slows down vehicles and impairs the comfort of drivers and passengers.

Saturation Flow is also considered to be the maximum number of vehicles crossing the intersection in the 1-hour period of the intersection's green time; Base Year: the year in which the traffic data was collected and the study was carried out; Year One: year in which the project is already up and running; Project Year: year in which the end of the project is calculated, i.e. the last year in which the project would work properly, in most years it is projected for 10 years; Project Hourly Volume (PHV): Volume adopted for dimensioning roads and intersections, levels of service, road operation planning, signaling, and traffic regulation; Level of Service: Measure of traffic quality on the road or intersection, which takes into account vehicle speed, comfort, ease of changing lanes, entering and exiting the road or intersection, and proximity of one vehicle to another. They are classified as

A, B, C, D, E and F (HCM, 2000).

At level A, the flow is free, with reduced concentration. The individual has total freedom in choosing the speed and total ease of overtaking (HCM, 2000), as shown in figure 1:

Figure 1: Level A: Comfort and convenience: Good.

Fonte: http://www.brazhuman.com.br/pdf/HCM-2000_412_aula_3_introd._operacao_rodoviaria.pdf.

At level B, the flow is stable, with reduced concentration. Freedom of choice of speed and ease of overtaking is not total, although still at a very good level. This is shown in figure 2.

Figure 2 - Level B: Comfort and convenience: Good.

Fonte: http://www.brazhuman.com.br/pdf/HCM-2000_412_aula_3_introd._operacao_rodoviaria.pdf.

Level C also has a stable flow, with a medium concentration. Freedom of choice of speed and ease of overtaking is relatively hampered by the presence of other vehicles, as shown in figure 3.

Figure 3 - Level C: Comfort and convenience: Fair.

Fonte: http://www.brazhuman.com.br/pdf/HCM-2000_412_aula_3_introd._operacao_rodoviaria.pdf.

Regarding level D, it can be said that it is close to the unstable flow. High concentration. Reduced freedom in the choice of speed and great difficulty in overtaking (HCM, 2000), as shown in figure 4.

Figure 4 - Level D: Comfort and convenience: Poor.

Fonte: http://www.brazhuman.com.br/pdf/HCM-2000_412_aula_3_introd._operacao_rodoviaria.pdf.

The flow can be extremely poor. This is related to level E, which shows unstable flow, with extremely high concentration. There is no freedom to choose speed and maneuvers to change lanes are only possible if forced. Figure 5 clearly represents this level.

Figure 5 - Level E: Comfort and convenience: Poor.

Fonte: http://www.brazhuman.com.br/pdf/HCM-2000_412_aula_3_introd._oper acao_rodoviaria.pdf.

At level F, the flow is forced, with very high concentration. Very low speeds and frequent long stops. Maneuvers to change lanes are only possible if forced and with the cooperation of another driver (HCM, 2000). This can be seen in figure 6.

Figure 6 - Level F: Unacceptable.

Fonte: http://www.brazhuman.com.br/pdf/HCM-2000_412_aula_3_introd._oper acao_rodoviaria.pdf.

It is worth noting the concept of Service Volume, which is the number of vehicles associated with each service level. Capacity is generally equal to the service volume corresponding to level D, so it is not convenient for the road to operate at full capacity.

2.4 Simulators

When an intersection is not optimally planned, as well as a road network, sizing errors occur, causing reduced capacity, increased waiting times, reduced saturation flow and worsening service levels.

For proper planning, simulation *software* is used to propose possible solutions using previously collected quantitative and qualitative data, analyzing variables including traffic volume on approach, saturation flow, lost time, average speed, level of service and queue size.

2.4.1 *TransCAD*

TransCAD software is used as a GIS-T (Geographic Information System for Transport) platform, and is used to analyze, manipulate, store and visualize vehicle traffic data. The tool works with mathematical modeling, distributing vehicles on the network through vehicle counts at intersections made in the field, generating an adjusted Origin - Destination matrix (O-D matrix).

The model created by *TransCAD* allows you to add data such as delays and restrictions on conversions, one-way roads, roads with different levels, centroid connectors, intermodal terminals, road hierarchy and performance functions, transfer points and delay functions.

In transportation planning, elements such as distance, travel times, and Origin - Destination flows are essential for manipulating data, which is done through matrices. *TransCAD* therefore provides functions that create and modify them in order to adjust them.

TransCAD makes it possible to dimension the traffic flow of the entire network and analyze congestion points using traffic allocation models. O-D matrices are taken as input data, and each flow is generated based on the travel time or the impediment of alternative paths that could be used to travel from that origin. There are several traffic allocation methods in the *software,* the one used in the project is the All or Nothing (AON) method, where the matrix is generated considering the shortest route between the origin and the destination.

2.4.2 *AimSun*

The simulator was created based on four fundamentals: integration, modularity, performance and customized extensions. With regard to integration, as the program evolves, new knowledge is improved, requiring only an update. Reusing tasks and processes defines the modularity of the *software.* The development of *Aimsun* is parallel to the evolution of processors, i.e. a few years ago it wasn't possible to scale the entire city of New York, now computer technology and *software* make it possible. Extensibility is defined as the ease with which users can program customized extensions (HALLMAN, 2011).

The characteristic of the movements basically works by the *car-following* model, which is guided by parameters such as road characteristics, driver type, and the influence of adjacent lanes. *Car-following* is made up of acceleration and deceleration: if the road is clear, it will develop until it reaches the maximum speed allowed; if the road is congested or even with only one car ahead, deceleration is the limitation of following the car ahead of everyone else (BARCELÓ, 2010 *apud* HALLMAN, 2011).

In this way, each vehicle is treated individually and its characteristics are defined in the input data, remaining unchanged during the simulation.

The *software*'s simple interface helps to create a road network, making it easy to add and customize a road network, such as streets, intersections, their specific movements and traffic light times.

When allocating demand, *AimSun* offers two options: O-D matrices or traffic states. For demands made by O-D matrices, it is necessary to create zones and centroids which are intended to generate and attract vehicles if the road is planned to be two-way, otherwise the centroid will only generate or attract depending on the direction of the street. To differentiate between vehicles in the simulation, it is necessary to enter specific matrices for each type of vehicle, such as one matrix for light vehicles and another for heavy vehicles.

Figure 7 shows how the process of generating and attracting vehicles by centroid takes place.

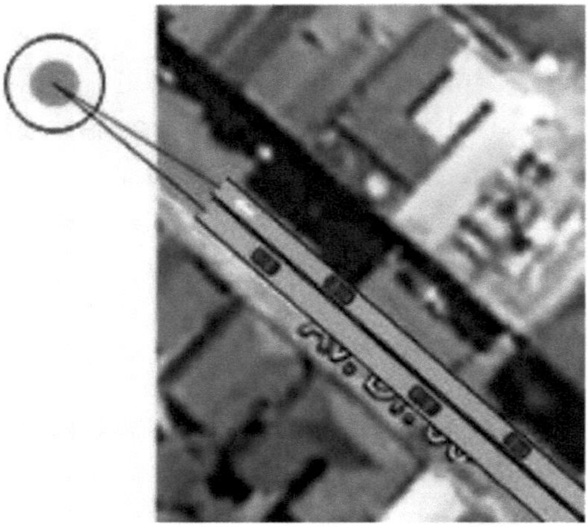

Figure 7 - Example of a hub generating and attracting vehicles

Source: Project drawn up by the author.

AimSun also allows you to import matrices from other *software* such as *Microsoft Excel.* This portability makes projects easier and optimizes simulation creation time.

Traffic states act as injectors and collectors of traffic from other traffic programs, but are not recommended for complex road networks (HALLMAN, 2011).

At the end of the network dimensioning, tools can be used to correct the road network, such as checking whether all the roads are interconnected or whether an O-D pair can have a path or not.

2.4.3 *SIDRA Intersection*

Among the *software* available, *SIDRA Intersection* was also used, which makes it possible to model various configurations, including: traffic light intersections; controlled non-safety intersections; traffic circles; traffic analysis of movements from secondary roads to main roads and from main roads to secondary roads; uninterrupted traffic flows.

In addition to *SIDRA* being an analysis tool for single crosses, the *software* can also be used to analyze coordinated crosses. *SIDRA* is a popular program and is generally used in North America, Europe and Oceania.

According to HCM 2000, in order to find the level of service of an intersection, it is necessary to calculate the vehicular delay (Delay), which is also used in the *SIDRA software*. The equations below show how to calculate vehicular delay, capacity, saturation flow, cycle time, effective green, total lost time, occupancy, optimum cycle and degree of saturation using Webster's formulas. Saturation flow is determined by equation 1.

$$S = 525.\ L \qquad (1)$$

In which:

S - Saturation flow;

L - Approach width, for $L > 5.50$ m and $L < 18.0$ m.

Green time is not fully utilized due to drivers' reactions during the start of green and the start of yellow. The lost time is then defined as the difference between the effective green time and the sum of the green and yellow times, according to equation 2.

$$g_{ef} = g + A - I \qquad (2)$$

In which

gef - Effective green time;

A - Yellow time;

I - Time lost per phase.

The cycle time is given by equation 3:

$$c = A + v + g \tag{3}$$

c - Cycle time

A - Yellow time;

V - Red time;

g - Green time.

The following equation 4 is used to calculate capacity:

$$C = \frac{S \cdot gef}{c} \tag{4}$$

C - Capacity;

S - Saturation Flow; gef - Effective Green Time; c - Cycle Time.

The degree of saturation is calculated using equation 5:

$$x_i = \frac{demanda}{Capacidade} = \frac{q_i}{\frac{S_i \cdot g_{ef}}{c}} \tag{5}$$

In which

Xi - Degree of saturation;

qi - Demand (volume);

Si - Saturation flow;

Gef - Effective green time;

c - Cycle time.

Vehicle Delay is calculated using equation 6:

$$d = \frac{c(1-\lambda)^2}{2(1-\lambda x)} + \frac{x^2}{2q(1-x)} - 0,65 \left(\frac{C}{q^2}\right)^{1/3} \cdot \quad x^{(2+5\lambda)} \tag{6}$$

In which

d - Average delay;

c - Cycle;

λ - Effective green ratio / cycle time;

q - Demand;

x - Degree of saturation;

C - Capacity.

To calculate the lost time (T_P), equation 7 is used.

$$T_P = \sum_{1}^{n} I_i \qquad n = n° \text{ de fases} \tag{7}$$

In which

Tp - Lost time;

I - Time lost per phase.

Equation 8 is used to calculate the occupancy rate (y).

$$y = \frac{q_i}{S^i} \tag{8}$$

y - Occupancy rate;

qi - Demand;

Si - Saturation flow.

Another equation is 9, the optimum cycle time (Co).

$$Co = \frac{1{,}5 \, . \, T_P + 5}{1 - Y} \tag{9}$$

In which

Co - Optimum cycle time;

Tp - total lost time;

Y - Sum of all occupancy rates.

The SIDRA *software was* used to analyze the level of service and was observed only through the delay. Table 1 shows the delay indices used to obtain the level of service generated:

Table 1 - Method of defining service level by delay alone.

Service level	**Delay per vehicle in seconds (d)**

A	$d \leq 10$
B	$10 < d \leq 20$
C	$20 < d \leq 35$
D	$35 < d \leq 55$
E	$55 < d \leq 80$
F	$d > 80$

Source: Prepared by the author, using knowledge from the SIDRA Intersection 5.1 user manual.

3 METHOD

Defining a solution for a given intersection and dimensioning its lanes depends on the volume and characteristics of the traffic that will circulate in the design year. It is necessary to obtain the Project Hourly Volumes (PHV), which must be represented in flowcharts of the various vehicle movements, classified as cars or trolleys, buses and trucks. The flowchart is made for the project year. Whenever possible, it is recommended that the Project Hourly Volume (PHV) is also expressed in passenger car units per hour (PCU/hour). It is common to use equivalence factors for buses, trucks and trailers in order to unify all vehicles into UCP/hour (BRASIL, 2005, p. 45).

According to Brasil (2005), to obtain the data needed to draw up the flowcharts, traffic counts were carried out at the intersections according to the following guidelines:

- Through specific observation of the site, determine the peak traffic periods. It is essential to identify the days of the week and time periods when peak traffic occurs.
- Identify the characteristics of the larger vehicles that use the intersection.
- Counting vehicles during peak periods and totaling every 15 minutes.

Counts will be made separately for the volumes of passenger cars, trucks, buses and trailers, according to the classification adopted.

- Determine whether or not there are seasonal factors for adjusting the results of the counts to the period of the year with the highest traffic flow.
- Use a survey of vehicle growth and increase rates over the last few years to apply to the volumes determined in the counts, obtained from existing socio-economic studies or traffic studies.

The study in question was filmed using a Sony DCR-PJ5 camera, as there was not enough manpower to carry out the counts in the field. Two peak periods were pre-established, by continuous observation, at the main intersection (referred to in the study as intersection 1) and the surrounding intersections (intersections 2, 3, 4 and 5) to form a network. The times established were: Morning peak: from 7am to 10am; and Afternoon peak: from 5pm to 8pm. Figure 8 shows the floor plan of the intersections studied.

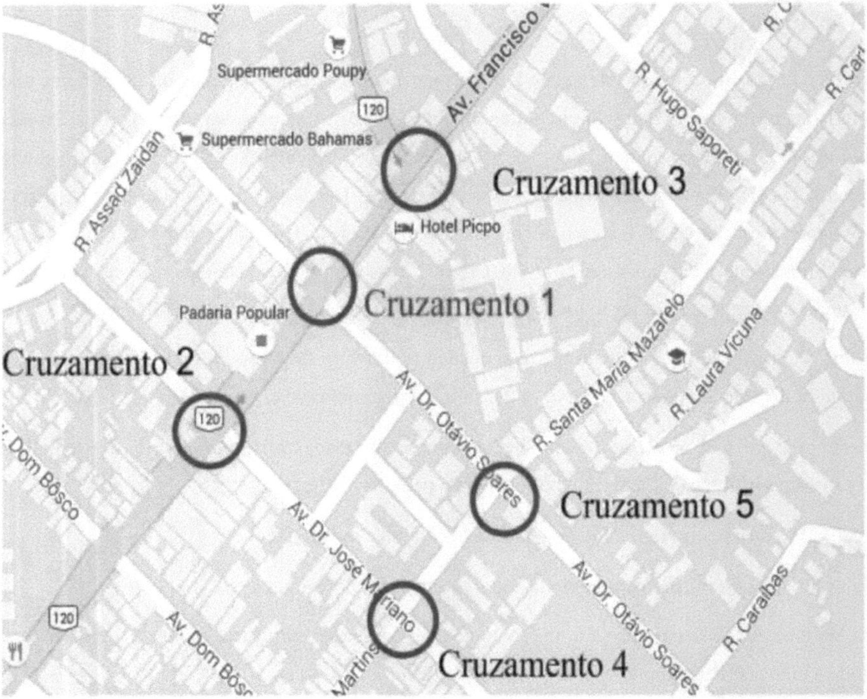

Figure 8 - Crossroads and Surroundings.

Source: Image taken from *Google Maps* and edited by the author.

After defining the road network and the intersections to be studied, it is necessary to identify all the possible movements at each intersection. To facilitate understanding, it was necessary to create sketches to show the direction and direction of the movements at each intersection and to name them so that the vehicle count table could be used later.

Junction 1 is using 2 traffic light phases, and when the phase allows red movements to pass, green movements are not possible, the same happens when the signal opens for green movements, preventing red movements from being executed. This is how it was stipulated:

- **Phase 1**: Movements 1A, 1B, 1C and 1G.

Traffic light timing:

Cycle time: 107 seconds.

Green - 47 seconds.

Yellow - 2 seconds.

Red - 58 seconds.

General red for this phase - 7 seconds.

- **Phase 2**: Movements 1D, 1E and 1F.

Traffic light timing:

Cycle time: 107 seconds.

Green - 42 seconds.

Yellow - 2 seconds.

Red - 63 seconds.

General red for this phase - 7 seconds.

The flowchart in figure 9 shows the movements in their respective phases:

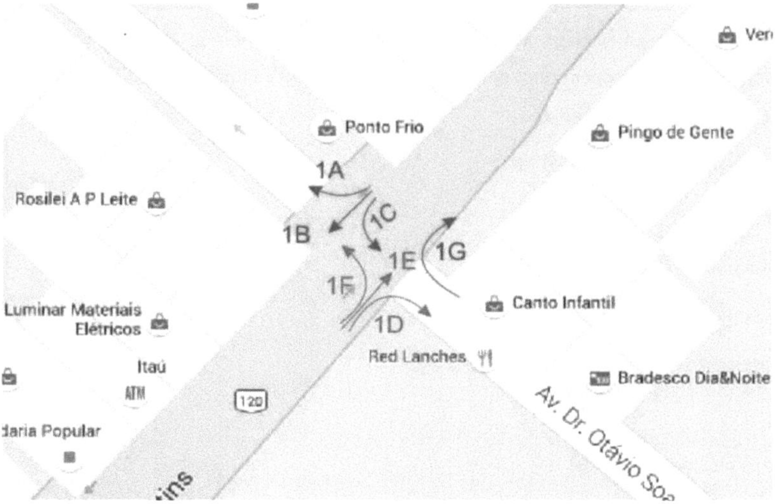

Figure 9 - Sketch of the permitted movements at Junction 1.

Source: Image taken from *Google Maps* and edited by the author.

In figure 10, it is still possible to see that *there are* trestles preventing vehicles coming from the Dr. Otâvio Soares Avenue approach from moving in a straight direction.

Figure 10 - Ground level view of junction 1.

Source: *Google Earth Street View.*

Junction 2 consists of 10 movements and 4 traffic lights, 2 of which use the same phase. The blue movements take part in the same phase, after the traffic lights for the blue movements close, the signal for the orange movements opens, and then after the orange movements finish, the traffic light opens for the pink movements. It is called:

- **Phase 1**: Movements 2A, 2B, 2G and 2F.

Traffic light timing:

Cycle time: 126 seconds.

Green - 29 seconds.

Yellow - 2 seconds.

Red - 95 seconds.

General red for this phase - 7 seconds.

- **Phase 2**: Movements 2C, 2D and 2E.

Traffic light timing:

Cycle time: 126 seconds.

Green - 40 seconds.

Yellow - 2 seconds.

Red - 84 seconds.

General red for this phase - 7 seconds.

- **Phase 3**: Movements 2H, 2I and 2J.

Traffic light timing:

Cycle time: 126 seconds.

Green - 30 seconds.

Yellow - 2 seconds.

Red - 94 seconds.

General red for this phase - 7 seconds.

Figure 11 shows the movements in detail.

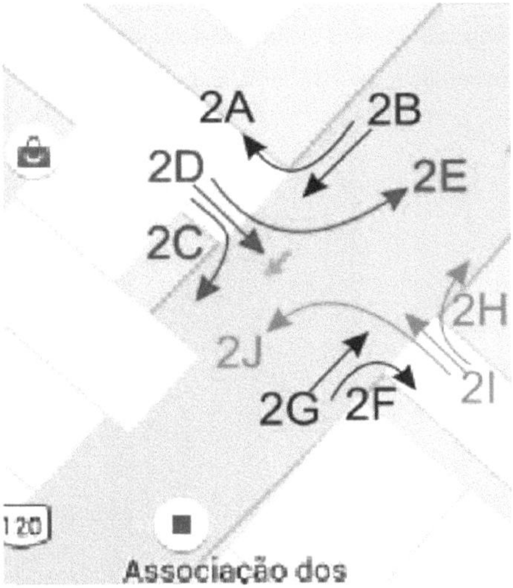

Figure 11 - Sketch of permitted movements at junction 2.

Source: Image taken from *Google Maps* and edited by the author.

Because the street where movements 2B, 2C and 2J end is narrower and has parking on both sides, it is forbidden for large cars to pass through. Figure 12 shows junction 2 and its intersecting roads.

Figure 12 - Ground level image of junction 2.

Source: *Google Street View*.

Intersection 3 is a road perpendicular to the city's main thoroughfare, with the only direction being towards Av. Francisco Vieira Martins. Two movements are allowed at the intersection: right turns and left turns. Figure 13 shows the intersection in plan and its respective movements.

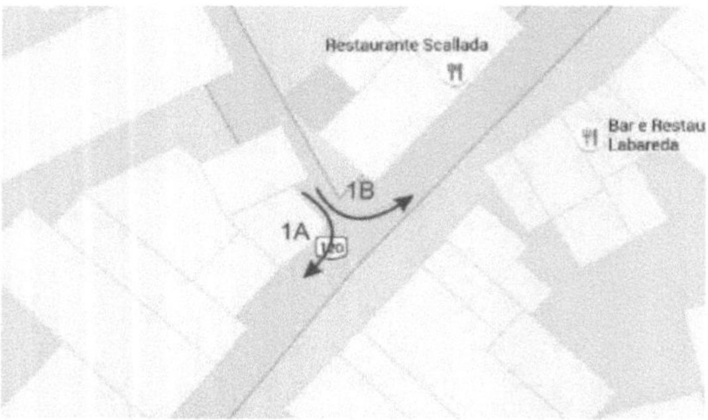

Figure 13 - Sketch of the permitted movements at junction 3. Source: Image taken from *Google Maps* and edited by the author.

According to the Brazilian Traffic Code (CTB), crossed yellow stripes at an intersection are defined as "conflict area markings", which aim to leave the defined space free for vehicles to enter the road. This marking is at junction 3, but because the horizontal signage is not up to date and perhaps due to recklessness on the part of drivers, vehicles are stopped on top of the lanes, obstructing the junction. And for this case, taking away the opportunity to turn left, turning right would be impossible because the traffic light at junction 1 would be closed and the congestion would reach junction 3. Figure 14 shows the current situation at junction 3.

Figure 14 - Ground level image of Junction 3.

Source: *Google Street View*.

Intersection 4 is made up of a small roundabout with 4 arrivals and 4 departures, i.e. all the streets are two-way and the intersection is made up of sixteen movements. Figure 15 shows the directions of the movements and their respective names used in this project.

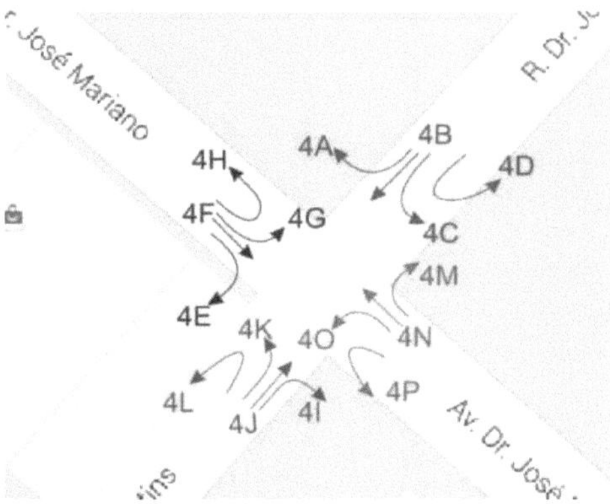

Figure 15 - Sketch of the permitted movements at Junction 4. Source: Image taken from *Google Maps* and edited by the author.

Although there is an island in the middle of the intersection, junction 4 operates with priority signs, with horizontal signs indicating the STOP sign on the approaches to Dr. José Mariano Avenue. In this way, the island has the function of reducing the speed of vehicles arriving on the approach, forcing left turns to be more open and facilitating U-turns, which are referred to at this junction as 4D, 4H, 4L and 4P. Figure 16 shows this in more detail.

Figure 16 - Ground level image of Crossroads 4.

Source: *Google Street View.*

The same applies to junction 5, a small diameter traffic circle in the middle of the intersection, with sixteen movements available. Figure 17 makes it easier to understand.

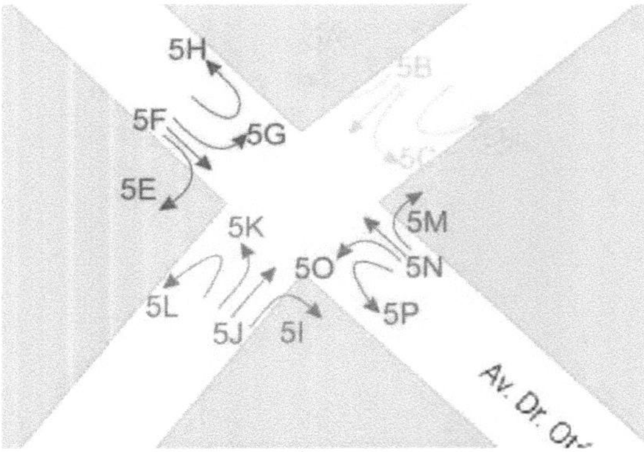

Figure 17 - Sketch of permitted movements at Junction 5.

Source: Image taken from *Google Maps* and edited by the author.

The ground level of Junction 5 is shown in Figure 18.

Figure 18 - Ground level image of Crossroads 5.

Source: *Google Street View*.

Once all the sketches had been drawn up, a simplified table was used, defining the location (intersection), the movement, the date, the hours divided every 15 minutes, and separated into light vehicles, buses, trucks and trailers. Table 2 shows the spreadsheet to be used for the counts.

Table 2 - Classified vehicle count used in the field

Classified Vehicle Count				
Location			Date	
Time:	Movement:			
	Light vehicles	Roads	Trailers	Buses

Source: Table prepared by the author.

After the counts were taken, they were entered into *Microsoft Office Excel* software in order to optimize the calculations to determine the hourly design volume for the morning and afternoon peaks, which will be extracted for the analysis of the intersection under study. Tables 3 and 4 show the sum of all the vehicles in an hour, at 15-minute intervals, using formulas developed in *Excel,* to determine the morning and afternoon peak hour, which will be used to calculate the corresponding volume for the project. It was not necessary to use any equivalence factor, as separate matrices will be made for light and heavy vehicles.

Table 3 - Final sum of morning peak hour volume

Time interval	Sum of vehicles

7am - 8am	3374
7:15- 8:15	3367
7:30-8:30	3477
7:45-8:45	3548
8am-9pm	3569
8:15-9:15	3516
8:30-9:30	3467
8:45 - 9:45	3291
9am-10am	2845

Source: Prepared by the author.

Table 4 - Final sum of afternoon peak hour volume

Time interval	Sum of vehicles
17h-18h	4470
17:15-18:15	4601
17:30-18:30	4597
17:45-18:45	4631
18:00 - 19:00	4629
18:15-19:15	4486
18:30-19:30	4305
18:45-19:45	3936
19h-20h	3185

Source: Prepared by the author.

It was therefore decided that the morning peak hour is from 8 am to 9 am and the afternoon peak hour is from 5:45 pm to 6:45 pm. From there, tables 5, 6, 7, 8 and 9 were assembled, summarizing the intersections and their respective movements in the morning and afternoon peak hours.

Table 5 - Summary of morning and afternoon peak hour flows at junction 1

Crossing 1	Morning (08:00 - 09:00)		Afternoon (17:45 - 18:45)	
	Lightweight	Heavy	Lightweight	Heavy
1A	280	27	334	25
1B	159	25	161	15
1C	132	29	208	21
1D	19	1	20	3

1E	218	49	298	34
1F	205	15	259	19
1G	94	6	185	5

Source: Prepared by the author.

Table 6 - Summary of Peak Hour Flows in the morning and afternoon at junction 2

Crossing 2	Morning (08:00 - 09:00)		Afternoon (17:45 - 18:45)	
	Lightweight	Heavy	Lightweight	Heavy
2A	86	24	70	17
2B	93	1	113	0
2C	38	0	42	0
2D	79	13	142	16
2E	112	55	151	43
2F	46	1	49	2
2G	231	8	221	8
2H	88	4	107	3
2I	125	24	95	23
2J	19	0	18	0

Source: Prepared by the author.

Table 7 - Summary of Peak Hour Flows in the morning and afternoon at junction 3

Crossing 3	Morning (08:00 - 09:00)		Afternoon (17:45 - 18:45)	
	Lightweight	Heavy	Lightweight	Heavy
3A	169	17	231	12
3B	55	8	78	1

Source: Prepared by the author.

Table 8 - Summary of Peak Hour Flows in the morning and afternoon at junction 4

Crossing 4	Morning (08:00 - 09:00)		Afternoon (17:45 - 18:45)	
	Lightweight	Heavy	Lightweight	Heavy
4A	42	14	70	15
4B	71	1	125	0

4C	22	2	35	1
4D	3	0	4	0
4E	40	5	45	0
4F	33	9	62	7
4G	48	2	46	3
4H	17	3	12	3
4I	26	4	39	1
4J	124	6	263	2
4K	58	4	79	1
4L	5	0	5	0
4M	25	1	31	0
4N	58	9	82	4
4O	24	1	17	0
4P	1	0	2	0

Source: Prepared by the author.

Table 9 - Summary of Peak Hour Flows in the morning and afternoon at junction 5

Crossing 5	Morning (08:00 - 09:00)		Afternoon (17:45 - 18:45)	
	Lightweight	Heavy	Lightweight	Heavy
5A	21	1	28	0
5B	61	3	55	1
5C	19	0	15	0
5D	1	0	3	0
5E	55	13	4	15
5F	30	0	55	1
5G	25	2	55	1
5H	9	0	16	0
5I	25	0	48	1
5J	79	0	156	1
5K	42	7	114	3
5L	7	0	4	0
5M	19	1	28	0
5N	48	1	44	2

5O	10	0	21	0
5P	1	0	21	0

Source: Prepared by the author.

After determining the peak hour and the volume of light and heavy vehicles, the network was set up in *TransCAD software, based* on the map of the site itself, taken from *Google Maps*. To create the network, it was necessary to generate the streets and their respective directions, create the possible movements at the intersections in question, create the start and end zones of the network and name them, enter variables such as the distance of each street and each movement, made using a scale calculated with the measurement of a street on the map with the measurement of a street made in the software, predetermined speed between 20 and 40 km/h depending on the type of movement, and the travel time for each street and the movements at the intersections. Figure 19 shows the network created for this project.

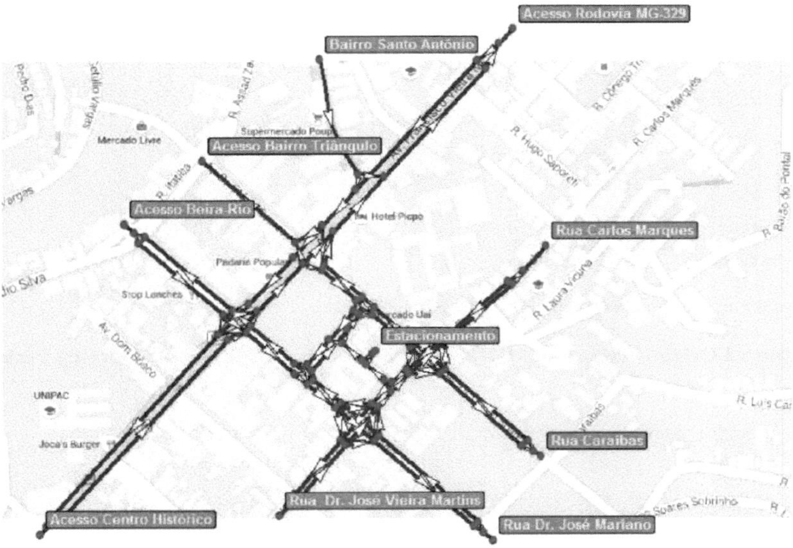

Figure 19 - Current Network drawn in TransCAD

Source: Photo taken from project by the author.

Once the network has been designed, the peak hour volumes will be added to their respective movements, and from there, through mathematical modeling carried out by the software itself, the vehicles will be distributed on the network, generating four Origin-Destination matrices. One for light vehicles and one for heavy vehicles in the morning, and the same for the afternoon. The same is done for heavy vehicles. This procedure is necessary because the counts were not done on the same day. The software presented an adjusted matrix, based on the data collected in the field, showing a better distribution in the network based on the data collected in the field. Finally, it will be possible to check the routes taken by the vehicles and identify the reason why there is congestion at the junction under study. Table 10 shows the O-D matrices generated and adjusted. These have been formatted in *Microsoft Excel* tables for better understanding.

Table 10 - Adjusted Origin-Destination Matrix of light vehicles in the morning rush hour

Light Manners	Bairro Triângulo access	Santo Antônio	Beira-Rio Access	Dr. José Mariano Street	Dr. José Vieira Mart	Caraibas Street	Access MG-329 highway	Access Historic Center	Carlos Marques Street	Parking	Total
Bairro Triângulo access	0	0	0	0	0	0	0	0	0	0	0
Santo Antônio neighborhood	41	0	32	11	13	12	55	34	8	17	224
Beira-Rio Access	40	0	0	27	13	7	61	38	5	34	226
Dr. José Mariano Street	24	0	28	0	24	0	30	5	0	0	112
Dr. José Vieira Mart Street	23	0	26	26	0	16	3	6	79	30	208
Caraibas Street	3	0	2	0	5	0	45	1	19	0	77
Access MG-329 highway	239	0	49	11	13	12	0	54	7	19	403
Access Historic Center	73	0	18	6	3	9	154	0	6	7	277
Carlos Marques Street	11	0	10	0	50	19	20	4	0	0	114
Parking	27	0	41	1	0	0	1	3	0	0	73
Total	483	0	206	83	121	76	370	145	124	107	1714

Source: Prepared by the author.

Table 11 shows the matrix of heavy vehicles in the morning peak.

Table 11 - Adjusted Origin-Destination Matrix of Heavy Goods Vehicles in the Morning Peak Hour

Heavy	Bairro Triângulo access	Santo Antônio	Beira-Rio Access	Dr. José Mariano Street	Dr. José Vieira Mart	Caraibas Street	Access MG-329 highway	Access Historic Center	Carlos Marques Street	Parking	Total
Bairro Triângulo access	0	0	0	0	0	0	0	0	0	0	0
Santo Antônio neighborhood	3	0	5	1	2	0	8	1	1	4	25
Beira-Rio Access	9	0	0	10	0	0	45	0	1	2	66
Dr. José Mariano Street	1	0	8	0	1	0	2	0	0	0	12
Dr. José Vieira Mart Street	1	0	2	4	0	0	1	0	0	6	14

Caraibas Street	0	0	0	0	0	0	1	0	1	0	2
Access MG-329 highway	24	0	19	1	4	0	0	0	1	10	58
Access Historic Center	3	0	2	0	0	0	5	0	0	0	9
Carlos Marques Street	1	0	3	0	0	1	1	0	0	0	7
Parking	1	0	9	0	0	0	1	0	0	0	11
Total	42	0	48	16	7	1	64	1	3	22	204

Source: Prepared by the author.

With regard to light vehicles, table 12 provides information on the afternoon peak.

Table 12 - Adjusted Origin-Destination Matrix of light vehicles in the afternoon peak hour

Afternoon Light	Bairro Triângulo	Santo Antônio	Beira-Rio Access	Dr. José Mariano	Dr. José Vieira Mart	Caraibas Street	Access MG-329	Access Historic Center	Carlos Marques Street	Parking	**Total**
Bairro Triângulo access	0	0	0	0	0	0	0	0	0	0	0
Santo Antônio neighborhood	57	0	31	18	4	22	78	44	20	35	309
Beira-Rio Access	51	0	0	59	35	6	102	42	6	34	336
Dr. José Mariano Street	41	0	34	0	17	0	49	6	0	0	148
Dr. José Vieira Mart Street	27	0	19	39	0	38	28	4	156	70	381
Caraibas Street	7	0	5	0	16	0	44	2	28	0	102
Access MG-329 highway	277	0	35	17	2	22	0	62	20	42	477
Access Historic Center	67	0	13	8	7	10	168	0	6	5	286
Carlos Marques Street	13	0	9	0	54	15	28	4	0	0	122
Parking	29	0	15	0	0	5	36	1	4	0	91
Total	569	0	161	142	136	119	534	166	240	186	2252

Source: Prepared by the author.

As for heavy vehicles in the afternoon peak, table 13 provides the information.

Table 13 - Adjusted Origin-Destination Matrix of Heavy Goods Vehicles in the afternoon peak hour

Afternoon Heavy	Bairro Triângulo access	Santo Antônio	Beira-Rio Access	Dr. José Mariano Street	Dr. José Vieira Mart	Caraibas Street	Access MG-329 highway	Access Historic Center	Carlos Marques Street	Parking	**Total**
Bairro Triângulo access	0	0	0	0	0	0	0	0	0	0	0
Santo Antônio neighborhood	3	0	4	1	0	0	1	0	0	5	13
Beira-Rio Access	12	0	0	9	0	1	29	0	1	4	56
Dr. José Mariano Street	1	0	3	0	0	0	0	0	0	0	4
Dr. José Vieira Mart Street	0	0	0	1	0	1	0	0	1	0	4
Caraibas Street	0	0	0	0	0	0	2	0	0	0	2
Access MG-329 highway	22	0	12	0	0	0	0	0	0	14	49
Access Historic Center	3	0	2	0	0	0	4	0	0	0	11
Carlos Marques Street	1	0	1	0	0	1	0	0	0	0	3
Parking	1	0	16	0	0	0	3	0	0	0	20
Total	44	0	39	11	0	3	40	0	2	23	161

Source: Prepared by the author.

4 RESULTS OBTAINED

Once the matrices have been generated, *TransCAD* enters the network flows for each specific origin and destination. Figures 20, 21, 22 and 23 show the networks with their respective flows added.

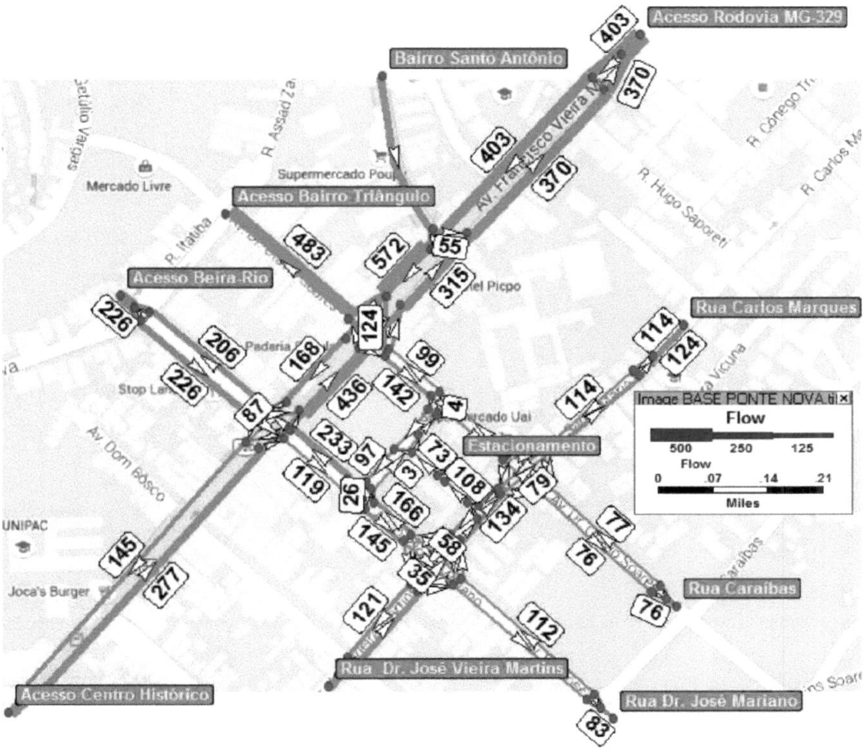

Figure 20 - Current network with the inclusion of the matrix of light vehicles in the morning rush hour generated by *TransCAD*.

Source: Photo taken from a project made by the author.

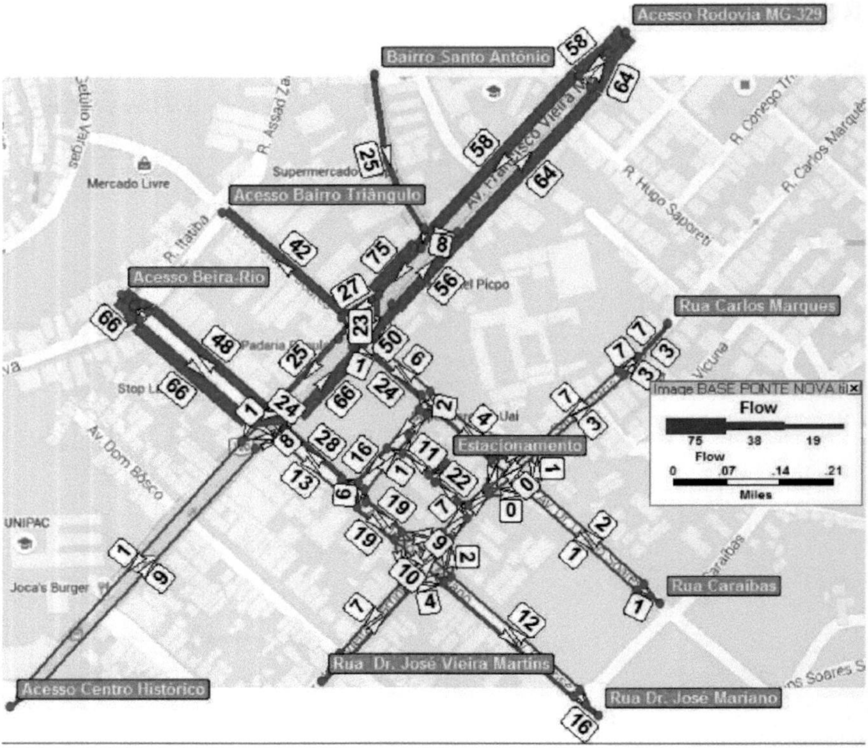

Figura 21 - Current network with the inclusion of the matrix of heavy vehicles in the morning rush hour generated by *TransCAD*.

Source: Photo taken from a project made by the author.

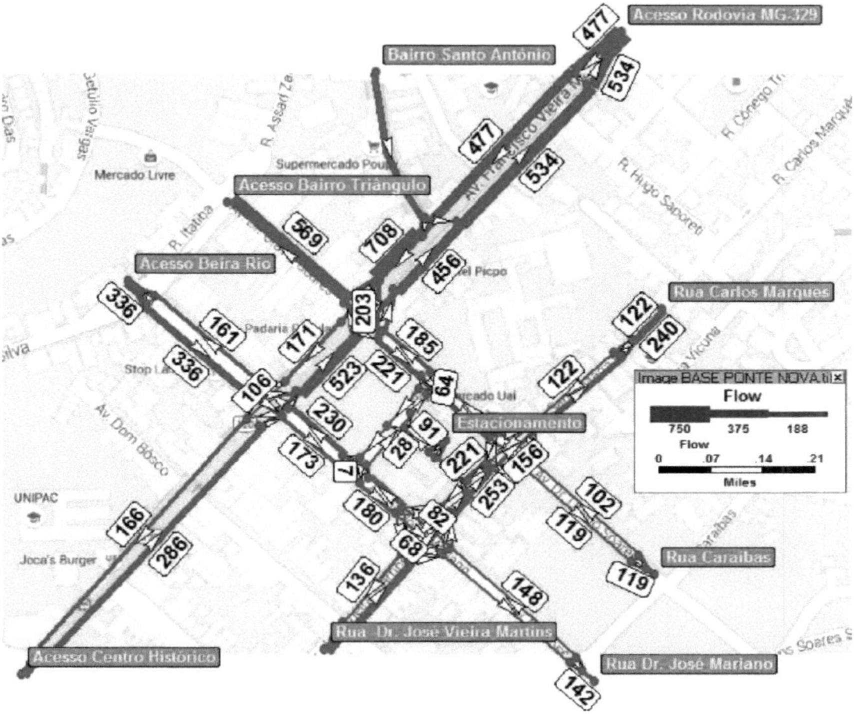

Figura 22 - Current network with the inclusion of the matrix of light vehicles in the afternoon peak hour generated by *TransCAD.*

Source: Photo taken from a project made by the author.

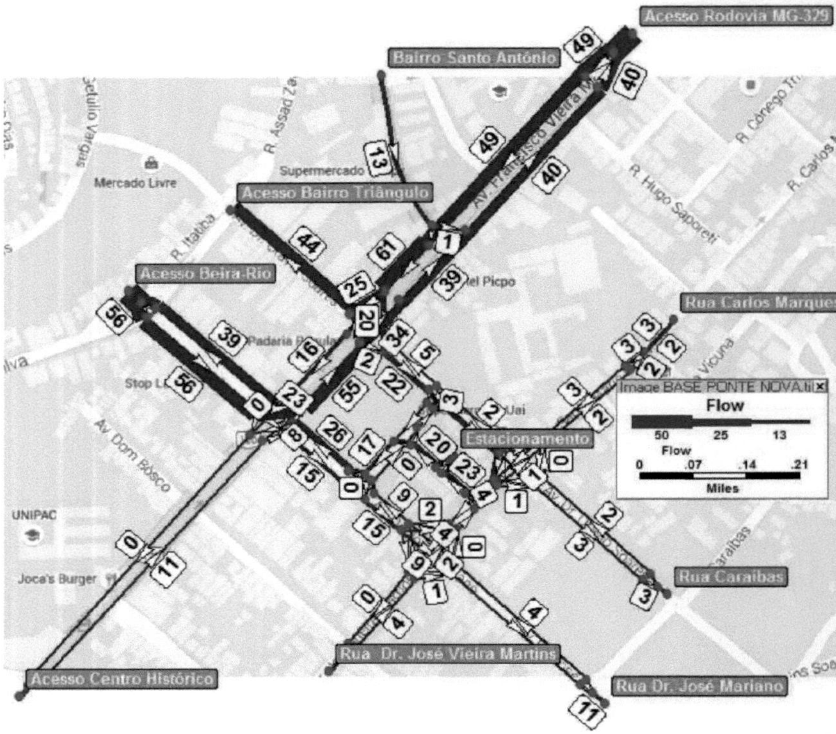

Figure 23 - Current network with the inclusion of the heavy vehicle matrix in the afternoon peak hour generated by *TransCAD*.

Source: Photo taken from a project made by the author.

The *software* uses street line thickness and flow numbers as parameters to visualize the number of vehicles circulating in a given region, making it easier to identify places where there is congestion.

After obtaining the O-D matrices for the two peak hour periods and for different vehicle sizes, it was observed that the afternoon peak hour had higher volume and consequently greater congestion. Because of this, in order to better illustrate the project, the light and heavy vehicle matrices for the afternoon period were simulated using the *AimSun software* to get a better view of the congestion on the roads. To do this, it was necessary to create a network identical to the one in *TransCAD,* create the streets and define their directions, the signs, the vehicle generation and attraction centroids, define the vehicle speed for all movements, and enter the cycle times and green, red and yellow times for each phase of each intersection. Finally, the afternoon peak hour matrices were imported into the simulator. Figure 24 shows different periods of the simulation.

Figure 24 - Simulation by AimSun software at junctions 1 and 3.

Source: Photo taken from project by the author.

It can be seen that the queue for movements 1A, 1B AND 1C is long, preventing movement 3A and congesting the arrival lane at the intersection. Although the *software* uses the driver's respect for the "conflict area marking" as a parameter, vehicles stop on top of the lanes, also hindering movement 3B, which could be executed if there are no vehicles crossing the avenue in the opposite direction, as shown in figure 25.

Figure 25 - *AimSun* software simulation of intersection 2.

Source: Photo taken from project by the author.

At junction 2, not only are the queues on the approaches relatively long, but there are also traffic jams with vehicles stopped in the middle of the intersection, preventing movements 2I, 2J and 2H, where their phase is open at the time the image was captured. The problem stems from intersection 1, where while phase 2 of the intersection is still closed, vehicles accumulate until they block intersection 2.

Since junction 1 is the main object of the study and is in the worst condition compared to the others, it was designed in the *SIDRA Intersection software,* and the movements currently allowed at the junction had to be defined. Figure 26 shows the program's sketch of the intersection in question.

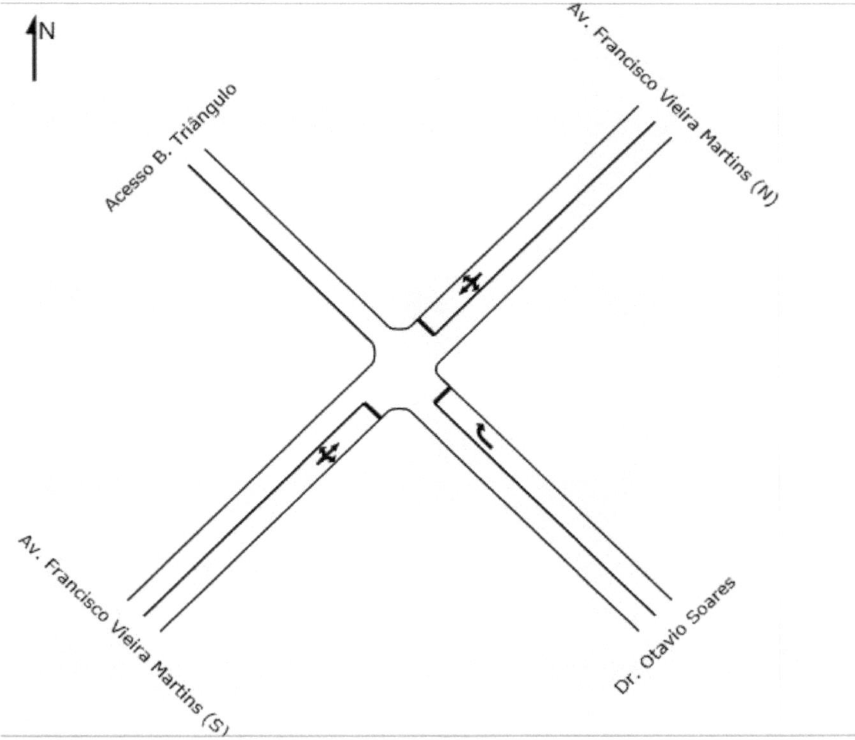

Figure 26 - *Layout of intersection* 1 using *SIDRA Intesection software.*

Source: Photo taken from project by the author.

In order to obtain the morning and afternoon service levels, it was necessary to enter into the *software* the speed at which vehicles arrive and leave the intersection (the same as used in *TransCAD*), the traffic light time for each phase and the volume of light and heavy vehicles in each movement. Other parameters used in *SIDRA* were kept as standard, and *Delay (*Vehicle Del*ay*) was used as the program's calculation method, which is proposed by HCM 2000. Figures 27 and 28 show the results for the morning and afternoon periods.

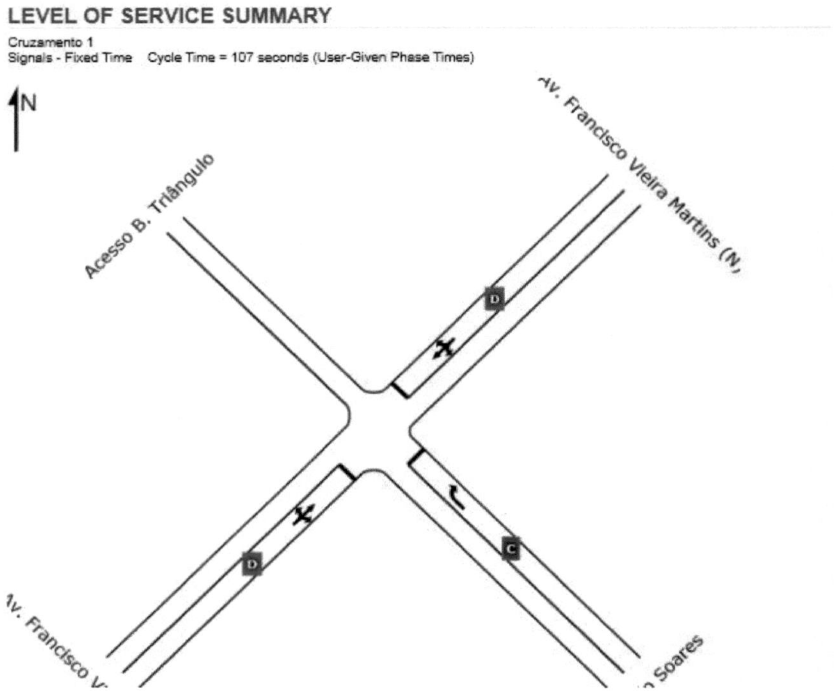

Figure 27 - Level of service of junction 1 obtained by *SIDRA Intersection software* in the morning.

Source: Photo taken from project by the author.

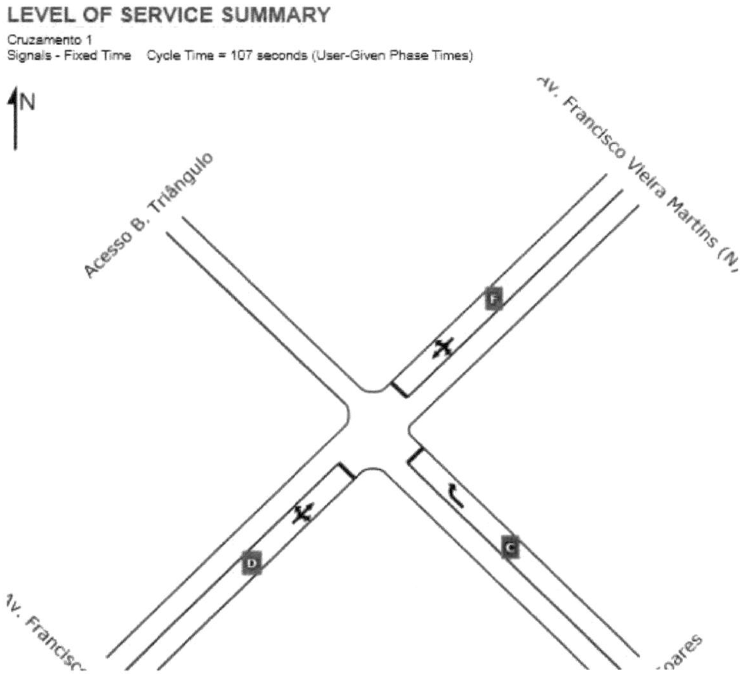

Figure 28 - Service level of junction 1 obtained by *Sidra Intersection software in* the afternoon

Source: Photo taken from project by the author.

It can be seen that the Francisco Vieira Martins Avenue (N) approach has been classified with level of service F, which makes it unacceptable for the traffic quality established by the *High Capacity Manual* (HCM).

4.1 Solution proposal

In order to avoid the long-term idleness of a major investment and the violation of signalization due to oversizing, the design year to be considered for intersections should not exceed the 10th year of useful life. This measure will allow each intersection to be used safely in its first years of operation.

Therefore, with a possible solution in mind, a projection rate for vehicle growth in the municipality of Ponte Nova - MG was calculated in *Excel.* Fleet data for the years 2011, 2012, 2013 and 2014 was previously collected in order to observe a possible percentage of vehicle growth each year, and to use it as a parameter for the adoption of the annual rate. Table 14 shows the analysis.

Table 14 - Annual growth rate to be used

Year	Total vehicles	% Growth
2011	21605	
2012	23498	8%
2013	25171	7%
2014	26435	5%
Annual rate used		4%

Source: Prepared by the author.

The annual rate was chosen on the basis that the percentage of growth becomes smaller each year, so it was estimated that the rate from 2014 to 2015 would be 4%. This figure can still be considered high when compared to the growth of other municipalities. After defining the rate, calculations were made to obtain an adjustment factor for 2025, which was multiplied by all the traffic volumes found in the O-D matrices. Table 15 shows the adjustment factor found for the year 2025. Tables 16, 17, 18 and 19 show the results of the matrices multiplied by the factor in question.

Table 15 - Growth adjustment factor to be used

Year	Fee	Fitting actor
2015	4,0%	1,0000
2016	4,0%	1,04
2017	4,0%	1,0816
2018	4,0%	1,124864
2019	4,0%	1,16985856
2020	4,0%	1,216652902
2021	4,0%	1,265319018
2022	4,0%	1,315931779
2023	4,0%	1,36856905
2024	4,0%	1,423311812
2025	4,0%	1,480244285

Source: Prepared by the author.

Table 16 - Origin-destination matrix for light vehicles during the morning, plus the adjustment factor.

Manha Leves - 2025	Bairro Triângulo	Santo Antônio	Beira-Rio Access	Dr. José Mariano	Dr. José Vieira Mart	Caraibas Street	Access MG-329	Access Historic	Carlos Marques	Parking	**Total**

Bairro Triângulo access	0	0	0	0	0	0	0	0	0	0	0
Santo Antônio neighborhood	61	0	47	17	19	18	81	50	12	26	332
Beira-Rio Access	59	0	0	40	19	10	91	56	7	51	335
Dr. José Mariano Street	36	0	42	0	36	0	45	8	0	0	166
Dr. José Vieira Mart Street	34	0	39	38	0	23	4	8	117	44	308
Caraibas Street	5	0	4	0	7	0	67	2	28	0	113
Access MG-329 highway	354	0	72	16	19	18	0	80	11	28	597
Access Historic Center	109	0	26	9	5	13	229	0	9	10	410
Carlos Marques Street	16	0	15	0	74	28	29	6	0	0	168
Parking	40	0	60	1	0	0	1	4	0	0	107
Total	715	0	305	122	179	112	547	214	184	159	2537

Source: Prepared by the author.

Table 17 - Origin-destination matrix for heavy vehicles in the morning, plus adjustment factor.

Manha Pesados - 2025	Bairro Triângulo access	Santo Antônio	Beira-Rio Access	Dr. José Mariano Street	Dr. José Vieira Mart	Caraibas Street	Access MG-329 highway	Access Historic Center	Carlos Marques Street	Parking	**Total**
Bairro Triângulo access	0	0	0	0	0	0	0	0	0	0	0
Santo Antônio neighborhood	5	0	8	1	3	0	12	1	1	6	37
Beira-Rio Access	13	0	0	14	0	0	66	0	1	3	98
Dr. José Mariano Street	2	0	11	0	1	0	3	0	0	0	17
Dr. José Vieira Mart Street	1	0	3	6	0	0	2	0	0	9	21
Caraibas Street	0	0	0	0	0	0	1	0	1	0	3
Access MG-329 highway	35	0	28	2	5	0	0	1	1	14	86
Access Historic Center	4	0	2	0	0	0	7	0	0	0	14
Carlos Marques Street	2	0	5	0	0	1	1	0	0	0	10
Parking	1	0	14	0	0	0	2	0	0	0	16
Total	63	0	71	23	11	1	94	2	5	32	303

Source: Prepared by the author.

Table 18 - Origin-Destination matrix of light vehicles in the afternoon plus adjustment factor

Lightweight Afternoon - 2025	Bairro Triângulo	Santo Antônio	Beira-Rio Access	Dr. José Mariano	Dr. José Vieira Mart	Caraibas Street	Access MG-329	Access Historic	Carlos Marques	Parking	Total
Bairro Triângulo access	0	0	0	0	0	0	0	0	0	0	0
Santo Antônio neighborhood	84	0	45	27	6	33	115	65	30	51	457
Beira-Rio Access	75	0	0	88	52	9	151	62	8	51	497
Dr. José Mariano Street	61	0	51	0	25	0	72	10	0	0	219
Dr. José Vieira Mart Street	40	0	28	58	0	56	42	6	231	103	564
Caraibas Street	10	0	7	0	24	0	65	3	41	0	151
Access MG-329 highway	410	0	52	25	3	33	0	92	29	62	706
Access Historic Center	99	0	19	12	11	15	249	0	9	8	423
Carlos Marques Street	19	0	13	0	80	22	41	5	0	0	181
Parking	43	0	22	0	0	8	53	2	6	0	134
Total	842	0	238	210	201	176	790	245	355	276	3333

Source: Prepared by the author.

Table 19 - Origin-destination matrix for heavy vehicles in the afternoon, plus the adjustment factor.

Afternoon Heavy - 2025	Bairro Triângulo	Santo Antônio	Beira-Rio Access	Dr. José Mariano	Dr. José Vieira Mart	Caraibas Street	Access MG-329	Access Historic	Carlos Marques	Parking	Total
Bairro Triângulo access	0	0	0	0	0	0	0	0	0	0	0
Santo Antônio neighborhood	4	0	6	1	0	0	1	0	0	7	19
Beira-Rio Access	18	0	0	13	0	2	43	0	1	5	83
Dr. José Mariano Street	1	0	4	0	0	0	0	0	0	0	6
Dr. José Vieira Mart Street	0	0	1	1	0	1	0	0	1	0	6
Caraibas Street	0	0	0	0	0	0	3	0	0	0	3
Access MG-329 highway	33	0	18	1	0	0	0	0	0	21	72
Access Historic Center	5	0	4	0	0	0	6	0	1	0	16
Carlos Marques Street	1	0	2	0	0	1	0	0	0	0	4

Parking	2	0	24	0	0	0	4	0	0	0	29
Total	65	0	58	16	0	5	59	0	3	34	238

Source: Prepared by the author.

After finding the ideal adjustment factor and introducing it into the existing matrices, the network was redesigned as a proposed solution in the *AimSun software*, and the matrices already multiplied by the growth adjustment factor were imported into the program and the simulation was run.

The network was redesigned with a view to future low-cost implementation. As a proposal, the idea was to remove the parking space on Av. Francisco Vieira Martins between junction 3 and junction 1 in the southbound direction, making it a dual carriageway, along with the exit street from junction 1 towards the Triângulo neighborhood. Av. Dr. Otàvio Soares would become a one-way street in the eastbound direction, and the same would happen with Av. Dr. José Mariano (parallel to Av. Dr. Otàvio Soares), but in the opposite direction. The streets perpendicular to these would remain a two-way street with a one-way lane to serve as a change of direction for drivers.

As the streets have changed, so have the intersections. Junction 1 retains the two phases with 3 traffic lights, but its movements have undergone variations. The arrival from Av. Francisco Vieira Martins (North) will be prevented from using the left turn, so it can use the same phase as the approach from the same avenue in the South direction, while the arrival from Av. Dr. Otàvio Soares will have only one phase for its movements, which will be free to choose any direction. To make things easier to understand, a flowchart has been drawn up showing the proposed changes to the intersection, as shown in figure 29.

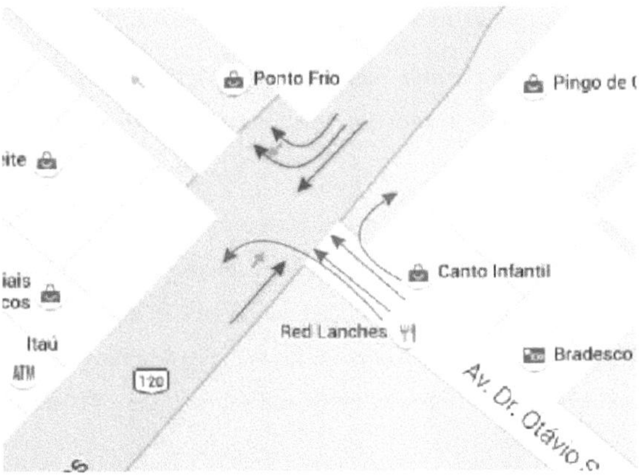

Figure 29 - Sketch of junction 1 with proposed changes Source: Photo taken from *Google Maps* and modified by the author.

In the flowchart, the movements in red participate in the same phase, and the movements in green use the other phase.

The same model was followed for junction 2, 3 traffic lights with two phases, with the arrivals from Av. Francisco Vieira Martins (north and south) sharing the same phase, while the next phase is used by the movements from the east arrival of Av. Dr. José Vieira Martins. Figure 30 shows the flowchart in detail.

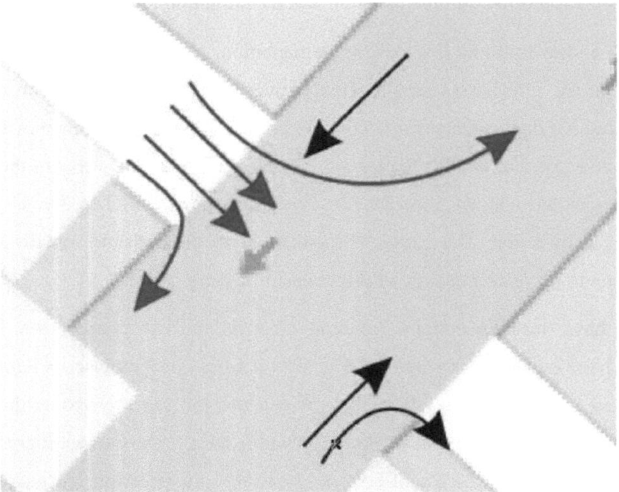

Figure 30 - Sketch of junction 2 with proposed changes.

Source: Photo taken from *Google Maps* and modified by the author.

Like the crossroads, movements in pink identify a phase, movements in blue use the next phase.

Intersection 3 has undergone only a small change. A possible increase in the width of the road on arrival at the intersection was created to make it easier to turn left when possible. In this way, the flowchart remained the same.

Intersection 4 has been exempted from the use of a mini-roundabout, with priority indications as a system, in the same way as it currently works, but the movements at the intersection have been altered and the return movements to the same road have been excluded due to lack of demand. The approach to Av. Dr. José Mariano (west) will serve as the arrival point for vehicles and the exit will be eastbound. Rua Dr. José Vieira Martins will remain two-way north and south. Figure 31 shows how the intersection has been changed.

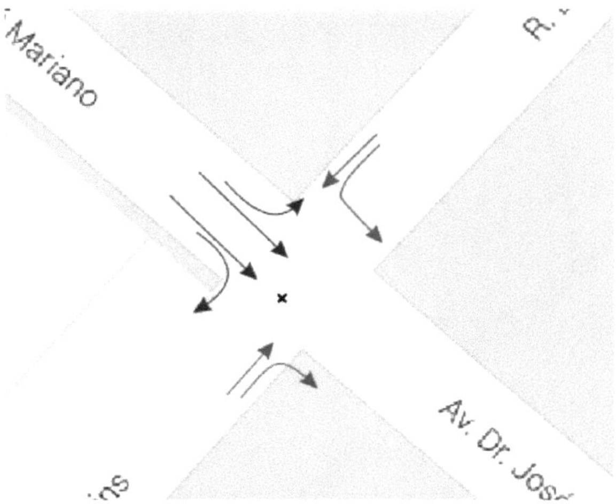

Figure 31 - Sketch of junction 4 with proposed changes.

Source: Photo taken from *Google Maps* and modified by the author.

The same applies to junction 5, but with the direction of Av. Dr. Otàvio Soares reversed, with vehicles arriving on the east approach and leaving on the west approach. Dr. José Vieira Martins and Santa Maria Mazarelo streets remain two-way. Figure 32 illustrates the changes.

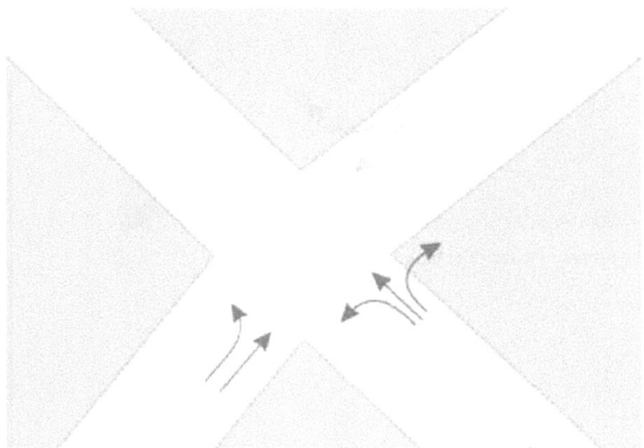

Figure 32 - Sketch of junction 5 with proposed changes.

Source: Photo taken from *Google Maps* and modified by the author.

Figures 33 and 34 illustrate the changes in the network and depict a short period of time during the simulation.

Figure 33 - *AimSun software* simulation of the proposed network at junctions 1, 2 and 3.

Source: Photo taken from project by the author.

Figure 34 - *AimSun* software simulation of the proposed network at junctions 4 and 5.

Source: Photo taken from project by the author.

You can see that the queue sizes have been considerably reduced at all intersections, significantly improving traffic flow. To prove the improvement, it was necessary to redesign the network in *TransCAD in the* same way as in *AimSun* and export the O-D matrices of light and heavy vehicles in the morning and afternoon. Figures 35, 36, 37 and 38 show the matrices relocated in the proposed network.

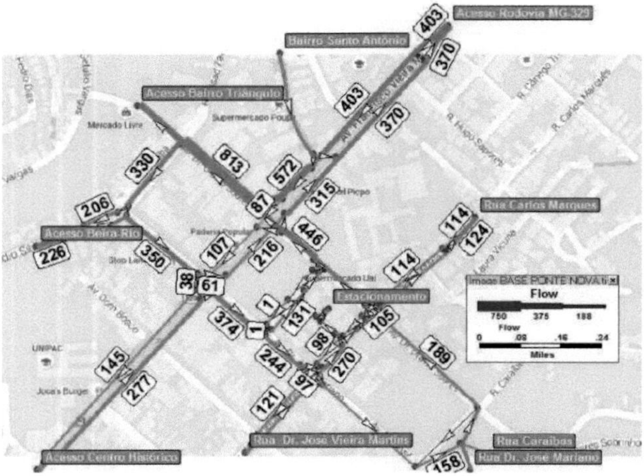

Figura 35 - Proposed network with the inclusion of the matrix of light vehicles in the morning rush hour generated by *TransCAD.*

Source: Photo taken from project by the author.

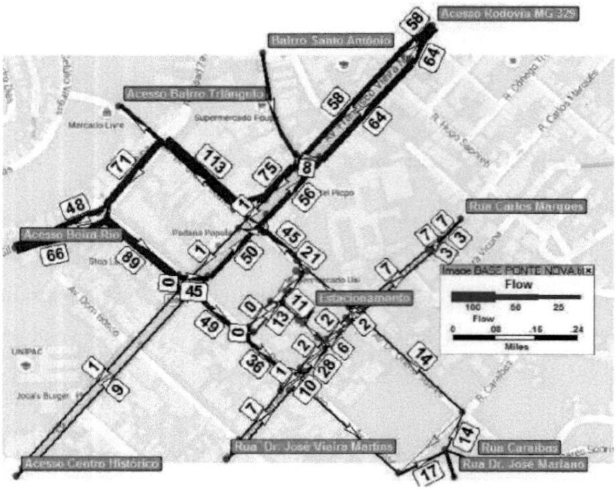

Figura 36 - Proposed network with the inclusion of the matrix of heavy vehicles in the morning rush hour generated by *TransCAD.*

Source: Photo taken from project by the author.

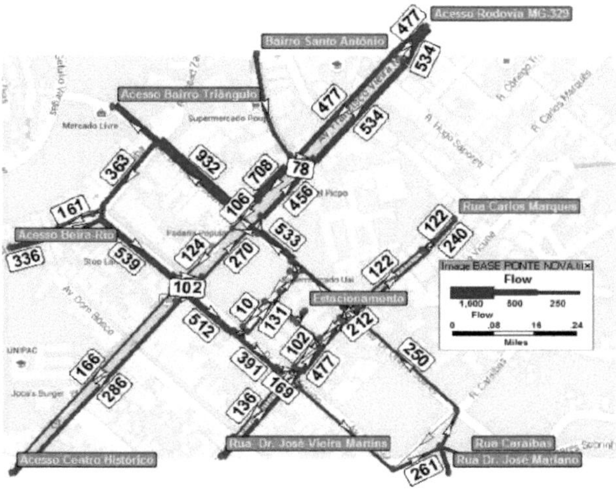

Figura 37 - Proposed network with the inclusion of the matrix of light vehicles in the afternoon peak hour generated by *TransCAD.*

Source: Photo taken from project by the author.

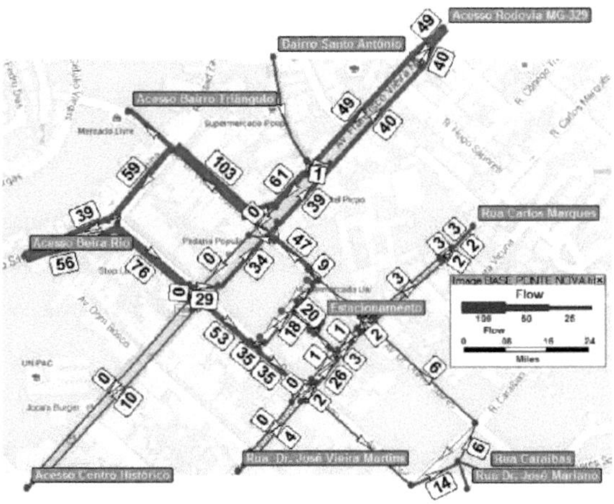

Figura 38 - Proposed network with the inclusion of the matrix of heavy vehicles in the afternoon peak hour generated by *TransCAD.*

Source: Photo taken from project by the author.

This procedure was necessary to find the volume of each new movement at each intersection of the proposed network, which will serve as input data to find the new Level of Service in force for intersection 1, which was redesigned in the *SIDRA Intersection software*. Figure 39 shows the sketch of junction 1 made by the *software.*

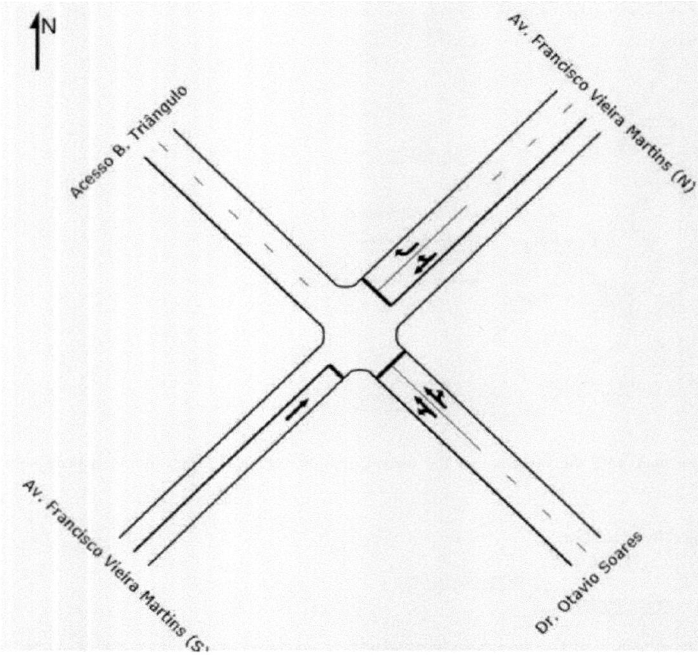

Figure 39 - *SIDRA Intersection* software drawing of proposed *intersection* 1.

Source: Photo taken from project by the author.

The other input data were standard ones already assigned by the program. The traffic light times were generated by the *software* using the optimal cycle. The levels of service for the morning and afternoon periods were calculated for the present day and for 2025, plus the annual growth factor calculated earlier. Figures 40, 41, 42 and 43 show the results.

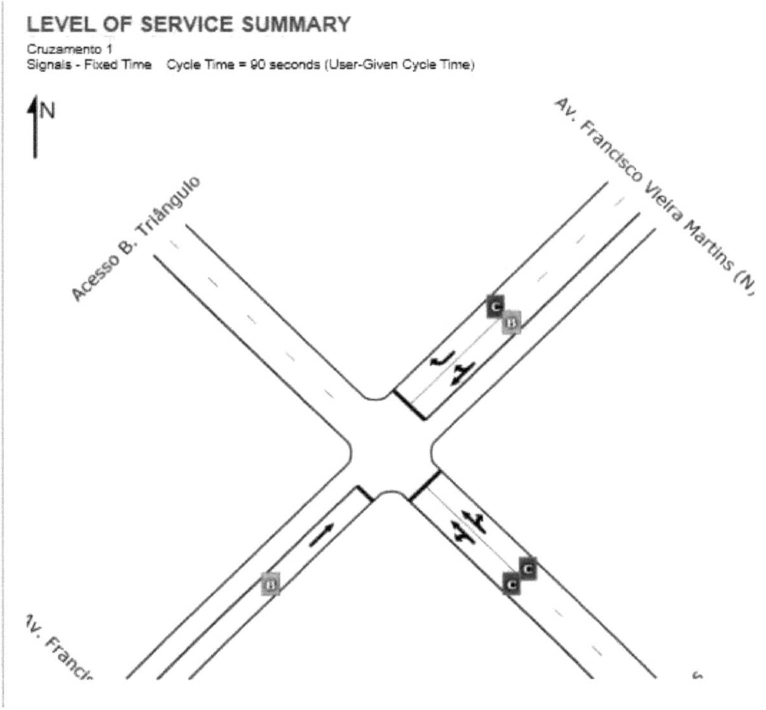

Figura 40 - Level of service in the morning these days for the proposed Crossing 1 made by the *SIDRA Intersection* software.

Source: Photo taken from project by the author.

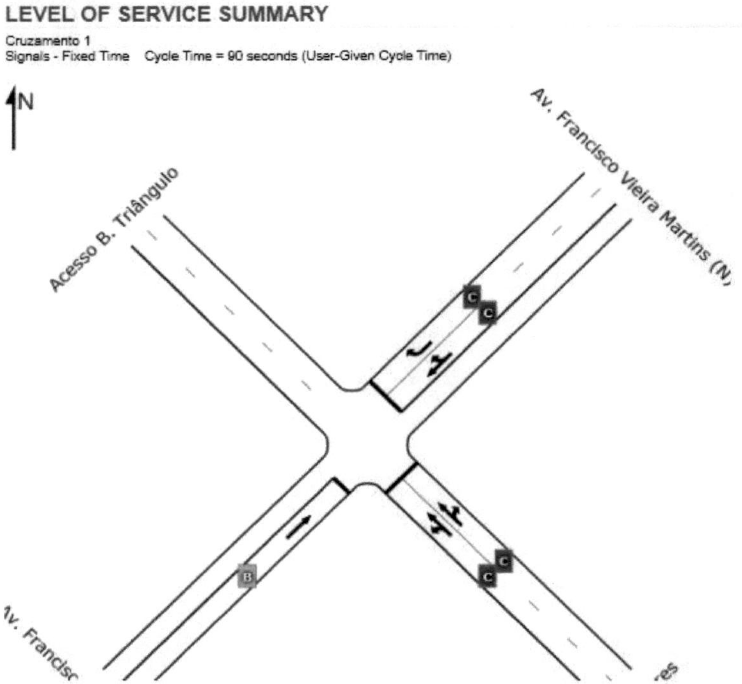

Figura 41 - Level of service in the afternoon these days for the proposed Crossing 1 made by the *SIDRA Intersection* software.

Photo taken from a project made by the author.

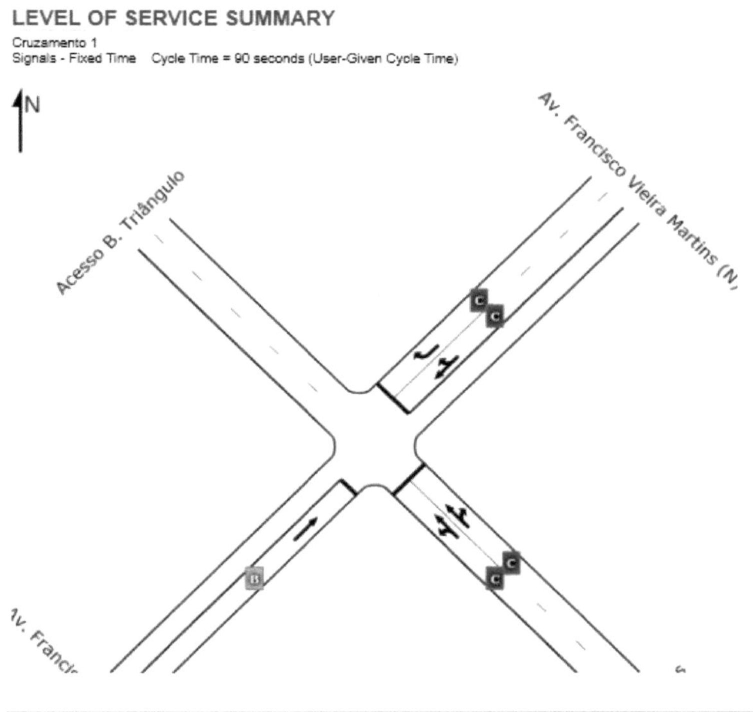

Figura 42 - Level of service in the morning in a projection for 2025 in the proposed Crossing 1 made by the *SIDRA Intersection* software.

Source: Photo taken from project by the author.

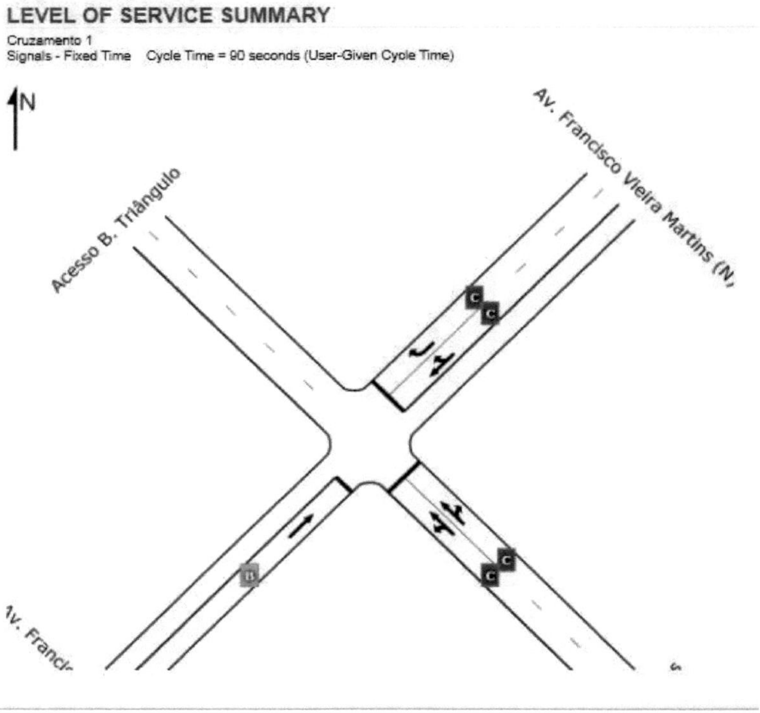

Figura 43 - Level of service in the afternoon in a projection for 2025 at the proposed Crossing 1 made by the *SIDRA Intersection* software.

Source: Photo taken from project by the author.

5 CONCLUSION

In this work we address the capacity analysis of intersections in the municipality of Ponte Nova - MG, made through vehicle counts, creation of the road network, and simulations made through specific *software.*

Based on the studies and data presented, it can be concluded that the proposed solution is fundamental for the municipality, and the level of service of the Francisco Vieira Martins Avenue (North) approach was rated F, which is considered unacceptable by the HCM. In the proposed solution, the minimum levels of service were C, which is defined as regular, bearing in mind that the project is for ten years later and a growth adjustment factor of approximately 1.48 was multiplied for each movement studied. This means that almost half of the vehicles already present on the current roads were added and yet intersection 1 performed better according to the SIDRA Intersection software.

In the simulations carried out by *AimSun*, you can clearly see the reduction in queues at intersections and the flow of vehicles was congestion-free throughout the network studied.

An interesting point about the project is that it aims to make changes that are economically viable for the municipality, and in the common conception, the cost of implementing these changes would simply be painting the new lanes and reorganizing the traffic signs according to the new directions and parking bans. But although it seems simple, it creates impasses between managers and the local community, which is why the change needs to be well evaluated by the competent bodies so that the majority of the population is satisfied with the change.

As a suggestion for future work, it was realized that it is possible to increase the road network used in the work, and add pedestrian traffic at intersections. Their influence can contribute to capacity and safety problems. Pedestrian movement counts should be carried out with a view to later analyzing the need to build pedestrian crossings or, possibly, to implement a special phase for pedestrians in the traffic light cycle. Based on this study, it would also be possible to carry out a capacity analysis of a road network for the municipality of Itabira - MG, where the UNIFEI Campus - Itabira facilities are located.

6 BIBLIOGRAPHICAL REFERENCES

BRAZIL. **Brazilian Traffic Code**. Ministry of Justice. Brasilia. 2002. Available at http://www.planalto.gov.br/ccivil_03/leis/l9503.htm. Accessed on 12 Oct 2015.

BRAZIL. **Traffic light manual.** Ministry of Justice. 1984. Available at http://meusite.mackenzie.com.br/professor_cucci/ManualSemaforos.pdf. Accessed on 12 Oct 2015.

BRAZIL. **Intersection design manual**. Ministry of Transport. 2005. Available at http://www1.dnit.gov.br/ipr_new/..%5Carquivos_internet%5Cipr%5Cipr_new%5Cmanua is%5CMANUAL_DE_PROJETO_DE_INTERSECOES_Versao_Final.pdf. Accessed on 12 Oct 2015.

CUCCI NETO, J. **Urban Traffic Engineering.** 2011. Available at http://meusite.mackenzie.br/professor_cucci/aula1.pdf. Accessed on 12 Oct 2015.

DEMARCHI, S. R.; SETTI, J. R. A. Capacity and Service Level Analysis of Highways *in* **Capacity and Service Level Analysis of Basic Highway Segments using HCM 2000.** Available at http://www.brazhuman.com.br/pdf/HCM- 2000_412_aula_3_introd._operacao_rodoviaria.pdf. Accessed on 11 Nov 2015.

FONTES (1995) *in* **Noções de Topografia Para Projetos Rodoviárias**. Chapter 4. Available at http://www.topografiageral.com/Curso/capitulo%2004.php. Accessed on 22 Oct 2015.

HCM - HIGHWAY CAPACITY MANUAL. **Capacity Studies - Introduction.** Available at http://www.dtt.ufpr.br/eng_trafego_optativa/arquivos/CAPACIDADE%20-%20INTRODUCTION.pdf. Accessed on 13 Oct 2015.

HALLMAN, H. V. **Comparison of Traffic Simulator *Software*.** 2011. Available at https://www.lume.ufrgs.br/bitstream/handle/10183/36891/000819131.pdf7seq uence=1. Accessed on 13 Oct 2015.

PIETRANTONIO; H.; BONETTI JR, W. **Use of Traffic-Actuated Traffic Lights.** 2002. Available at http://www.sinaldetransito.com.br/artigos/uso_e_parametrizacao_de_se maforos_atuados_%20pelo_trafego.pdf. Accessed on 12 Oct 2015.

VASCONCELLOS, E. A. **Circular é Preciso, Viver não é Preciso - A História do trànsito na Cidade de Sao Paulo.** FADESP. 1999. Available at https://books.google.com.br/books?id=rSyTuzbGi0C&printsec=frontcover&dq=inauthor:%22Eduardo+Alc%C3%A2ntara+de+V asconcellos%22&hl=enPT&sa=X&ved=0CBsQ6AEwAGoVChMI4K_y_LmGyQIVxCOQCh 2mLQ2V#v=onepage&q&f=false. Accessed on 12 Oct 2015.

7 APPENDIX

Appendix 1 - Vehicle flow tables

The tables represent the counts made in the field for each movement at their respective intersections, during the morning (07:00 to 10:00) and afternoon (17:00 to 20:00) peak hours.

Crossing: 1 Movement: A Date: 10/09/2015

Time		Lightweight	Trucks Trailers Buses			Heavy		Total
of	to		2 Axles	6 E	2 E	%	Total	
07:00	07:15	84	4	-	8	13%	12	96
07:15	07:30	86	3	-	2	5%	5	91
07:30	07:45	68	2	1	3	8%	6	74
07:45	08:00	97	8	-	6	13%	14	111
08:00	08:15	91	8	-	2	10%	10	101
08:15	08:30	61	3	-	1	6%	4	65
08:30	08:45	72	5	-	2	9%	7	79
08:45	09:00	56	5	-	1	10%	6	62
09:00	09:15	100	8	-	-	7%	8	108
09:15	09:30	80	-	-	2	2%	2	82
09:30	09:45	55	4	-	-	7%	4	59
09:45	10:00	69	4	-	-	5%	4	73
17:00	17:15	75	4	-	2	7%	6	81
17:15	17:30	71	2	-	1	4%	3	74
17:30	17:45	73	4	-	-	5%	4	77
17:45	18:00	93	3	-	-	3%	3	96
18:00	18:15	86	2	-	2	4%	4	90
18:15	18:30	88	3	-	4	7%	7	95
18:30	18:45	67	5	-	6	14%	11	78
18:45	19:00	86	3	-	4	8%	7	93
19:00	19:15	99	2	-	1	3%	3	102
19:15	19:30	82	3	-	2	6%	5	87
19:30	19:45	74	1	-	1	3%	2	76
19:45	20:00	77	3	-	-	4%	3	80
Total		1.890	-	1	50	3%	51	1.941

Crossing: 1 Movement: B Date: 10/09/2015

Time		Lightweight	Trucks Trailers Buses			Heavy		Total
of	to		2 Axles	6 E	2 E	%	Total	
07:00	07:15	34	2	5	-	17%	7	41
07:15	07:30	33	2	-	2	11%	4	37
07:30	07:45	33	1	1	-	6%	2	35
07:45	08:00	41	3	2	-	11%	5	46
08:00	08:15	33	4	2	-	15%	6	39
08:15	08:30	42	4	-	1	11%	5	47
08:30	08:45	42	6	-	1	14%	7	49
08:45	09:00	42	7	-	-	14%	7	49
09:00	09:15	18	3	2	1	25%	6	24
09:15	09:30	39	6	1	-	15%	7	46
09:30	09:45	41	4	2	-	13%	6	47
09:45	10:00	41	3	-	-	7%	3	44
17:00	17:15	35	7	-	-	17%	7	42
17:15	17:30	48	6	-	-	11%	6	54
17:30	17:45	52	9	-	-	15%	9	61

17:45	18:00	44	2	-	-	4%	2	46
18:00	18:15	48	5	-	1	11%	6	54
18:15	18:30	37	1	-	-	3%	1	38
18:30	18:45	32	4	1	1	16%	6	38
18:45	19:00	39	2	1	-	7%	3	42
19:00	19:15	59	2	-	1	5%	3	62
19:15	19:30	34	7	1	-	19%	8	42
19:30	19:45	34	2	2	-	11%	4	38
19:45	20:00	57	3	-	-	5%	3	60
Total		958	-	20	8	3%	28	986

Crossing: 1 Movement: C Date: 10/09/2015

Time		Lightweight	Trucks Trailers Buses			Heavy		Total
of	to		2 Axles	6 E	2 E	%	Total	
07:00	07:15	41	3	-	1	9%	4	45
07:15	07:30	35	3	-	3	15%	6	41
07:30	07:45	37	3	-	2	12%	5	42
07:45	08:00	42	1	-	3	9%	4	46
08:00	08:15	30	4	-	1	14%	5	35
08:15	08:30	28	5	1	1	20%	7	35
08:30	08:45	32	3	-	2	14%	5	37
08:45	09:00	42	7	-	5	22%	12	54
09:00	09:15	31	-	-	1	3%	1	32
09:15	09:30	40	3	-	2	11%	5	45
09:30	09:45	31	2	-	2	11%	4	35
09:45	10:00	41	1	-	1	5%	2	43
17:00	17:15	47	1	-	3	8%	4	51
17:15	17:30	51	-	-	3	6%	3	54
17:30	17:45	52	1	-	4	9%	5	57
17:45	18:00	45	2	-	4	12%	6	51
18:00	18:15	59	2	-	2	6%	4	63
18:15	18:30	65	1	-	6	10%	7	72
18:30	18:45	39	-	-	4	9%	4	43
18:45	19:00	52	1	-	4	9%	5	57
19:00	19:15	45	1	-	1	4%	2	47
19:15	19:30	40	-	-	3	7%	3	43
19:30	19:45	38	-	-	2	5%	2	40
19:45	20:00	34	-	-	1	3%	1	35
Total		997	-	1	61	6%	62	1.059

Crossing: 1 Movement: D Date: 10/09/2015

Time		Lightweight	Trucks Trailers Buses			Heavy		Total
of	to		2 Axles	6 E	2 E	%	Total	
07:00	07:15	2	-	-	-	0%	-	2
07:15	07:30	3	-	-	1	25%	1	4
07:30	07:45	4	-	-	-	0%	-	4
07:45	08:00	4	-	-	-	0%	-	4
08:00	08:15	6	-	-	1	14%	1	7
08:15	08:30	3	-	-	-	0%	-	3
08:30	08:45	7	-	-	-	0%	-	7
08:45	09:00	3	-	-	-	0%	-	3
09:00	09:15	7	-	-	-	0%	-	7
09:15	09:30	1	-	-	-	0%	-	1
09:30	09:45	1	-	-	-	0%	-	1
09:45	10:00	4	-	-	-	0%	-	4
17:00	17:15	2	-	-	-	0%	-	2
17:15	17:30	3	-	-	-	0%	-	3
17:30	17:45	6	-	-	-	0%	-	6

17:45	18:00	1	1	-	-	50%	1	2
18:00	18:15	7	-	-	-	0%	-	7
18:15	18:30	6	-	-	1	14%	1	7
18:30	18:45	6	1	-	-	14%	1	7
18:45	19:00	3	-	-	-	0%	-	3
19:00	19:15	10	-	-	-	0%	-	10
19:15	19:30	6	-	-	-	0%	-	6
19:30	19:45	1	1	-	-	50%	1	2
19:45	20:00	4	-	-	-	0%	-	4
Total		100	-	-	3	3%	3	103

Crossing: 1 Movement: E Date: 10/09/2015

Time		Lightweight	Trucks Trailers Buses			Heavy		Total
of	to		2 Axles	6 E	2 E	%	Total	
07:00	07:15	54	7	2	4	19%	13	67
07:15	07:30	42	7	1	2	19%	10	52
07:30	07:45	63	5	1	2	11%	8	71
07:45	08:00	44	5	4	6	25%	15	59
08:00	08:15	50	5	-	2	12%	7	57
08:15	08:30	62	11	1	3	19%	15	77
08:30	08:45	50	13	-	3	24%	16	66
08:45	09:00	56	9	-	2	16%	11	67
09:00	09:15	62	9	1	4	18%	14	76
09:15	09:30	58	6	-	1	11%	7	65
09:30	09:45	60	4	-	2	9%	6	66
09:45	10:00	59	12	-	2	19%	14	73
17:00	17:15	64	11	-	4	19%	15	79
17:15	17:30	74	7	2	3	14%	12	86
17:30	17:45	49	7	-	7	22%	14	63
17:45	18:00	64	2	1	3	9%	6	70
18:00	18:15	69	9	2	2	16%	13	82
18:15	18:30	76	-	1	3	5%	4	80
18:30	18:45	89	5	1	5	11%	11	100
18:45	19:00	67	5	-	5	13%	10	77
19:00	19:15	56	6	-	4	15%	10	66
19:15	19:30	64	7	-	4	15%	11	75
19:30	19:45	71	6	-	2	10%	8	79
19:45	20:00	75	4	2	1	9%	7	82
Total		1.478	-	19	76	6%	95	1.573

Crossing: 1 Movement: F Date: 10/09/2015

Time		Lightweight	Trucks Trailers Buses			Heavy		Total
of	to		2 Axles	6 E	2 E	%	Total	
07:00	07:15	51	2	-	2	7%	4	55
07:15	07:30	48	4	-	4	14%	8	56
07:30	07:45	46	1	-	4	10%	5	51
07:45	08:00	38	-	-	2	5%	2	40
08:00	08:15	50	1	-	4	9%	5	55
08:15	08:30	47	1	-	-	2%	1	48
08:30	08:45	53	3	-	2	9%	5	58
08:45	09:00	55	3	-	1	7%	4	59
09:00	09:15	57	4	-	3	11%	7	64
09:15	09:30	44	4	-	2	12%	6	50
09:30	09:45	49	2	-	-	4%	2	51
09:45	10:00	51	3	-	3	11%	6	57
17:00	17:15	63	4	-	4	11%	8	71
17:15	17:30	55	3	-	1	7%	4	59
17:30	17:45	55	1	-	2	5%	3	58

17:45	18:00	65	1	1	4	8%	6	71
18:00	18:15	73	2	-	3	6%	5	78
18:15	18:30	61	1	-	1	3%	2	63
18:30	18:45	60	-	1	2	5%	3	63
18:45	19:00	75	4	-	3	9%	7	82
19:00	19:15	51	3	1	1	9%	5	56
19:15	19:30	51	-	1	3	7%	4	55
19:30	19:45	45	-	-	1	2%	1	46
19:45	20:00	40	2	-	1	7%	3	43
Total		1.283	-	4	53	4%	57	1.340

Crossing: 1 Movement: G Date: 10/09/2015

Time		Lightweight	Trucks Trailers Buses			Heavy		Total
of	to		2 Axles	6 E	2 E	%	Total	
07:00	07:15	39	-	-	3	7%	3	42
07:15	07:30	16	1	-	-	6%	1	17
07:30	07:45	27	-	-	-	0%	-	27
07:45	08:00	43	-	-	-	0%	-	43
08:00	08:15	21	2	-	-	9%	2	23
08:15	08:30	22	1	-	2	12%	3	25
08:30	08:45	24	1	-	-	4%	1	25
08:45	09:00	27	-	-	-	0%	-	27
09:00	09:15	30	1	-	-	3%	1	31
09:15	09:30	23	1	-	-	4%	1	24
09:30	09:45	14	3	-	-	18%	3	17
09:45	10:00	37	1	-	-	3%	1	38
17:00	17:15	35	2	-	1	8%	3	38
17:15	17:30	40	-	-	2	5%	2	42
17:30	17:45	47	2	-	-	4%	2	49
17:45	18:00	43	-	-	-	0%	-	43
18:00	18:15	42	1	-	-	2%	1	43
18:15	18:30	43	3	-	1	9%	4	47
18:30	18:45	57	-	-	-	0%	-	57
18:45	19:00	62	-	-	-	0%	-	62
19:00	19:15	37	-	-	-	0%	-	37
19:15	19:30	32	2	-	-	6%	2	34
19:30	19:45	25	-	-	-	0%	-	25
19:45	20:00	31	-	-	-	0%	-	31
Total		817	-	-	9	1%	9	826

Crossing: 2 Movement: A Date: 15/09/2015

Time		Lightweight	Trucks Trailers Buses			Heavy		Total
of	to		2 Axles	6 E	2 E	%	Total	
07:00	07:15	27	7	1	-	23%	8	35
07:15	07:30	17	6	3	-	35%	9	26
07:30	07:45	19	1	-	-	5%	1	20
07:45	08:00	25	5	1	-	19%	6	31
08:00	08:15	23	4	-	-	15%	4	27
08:15	08:30	19	5	3	-	30%	8	27
08:30	08:45	18	5	1	-	25%	6	24
08:45	09:00	26	4	2	-	19%	6	32
09:00	09:15	21	4	1	-	19%	5	26
09:15	09:30	10	5	2	-	41%	7	17
09:30	09:45	23	3	1	-	15%	4	27
09:45	10:00	20	5	1	-	23%	6	26
17:00	17:15	17	8	-	-	32%	8	25
17:15	17:30	20	7	3	-	33%	10	30
17:30	17:45	25	3	-	-	11%	3	28

17:45	18:00	19	3	1	-	17%	4	23
18:00	18:15	15	1	1	-	12%	2	17
18:15	18:30	18	5	-	1	25%	6	24
18:30	18:45	18	5	-	-	22%	5	23
18:45	19:00	8	3	-	-	27%	3	11
19:00	19:15	13	-	1	3	24%	4	17
19:15	19:30	12	2	-	-	14%	2	14
19:30	19:45	8	2	3	-	38%	5	13
19:45	20:00	11	3	-	-	21%	3	14
Total		432	-	25	4	6%	29	461

Crossing: 2 Movement: B Date: 15/09/2015

Time		Lightweight	Trucks Trailers Buses			Heavy		Total
of	to		2 Axles	6 E	2 E	%	Total	
07:00	07:15	16	-	-	-	0%	-	16
07:15	07:30	13	-	-	-	0%	-	13
07:30	07:45	19	-	-	-	0%	-	19
07:45	08:00	35	-	-	-	0%	-	35
08:00	08:15	31	-	-	-	0%	-	31
08:15	08:30	19	-	-	-	0%	-	19
08:30	08:45	20	-	-	-	0%	-	20
08:45	09:00	23	-	-	-	0%	-	23
09:00	09:15	17	-	-	-	0%	-	17
09:15	09:30	13	-	-	-	0%	-	13
09:30	09:45	25	-	-	-	0%	-	25
09:45	10:00	22	-	-	-	0%	-	22
17:00	17:15	23	-	-	-	0%	-	23
17:15	17:30	30	-	-	-	0%	-	30
17:30	17:45	24	-	-	-	0%	-	24
17:45	18:00	16	-	-	-	0%	-	16
18:00	18:15	32	-	-	-	0%	-	32
18:15	18:30	32	-	-	-	0%	-	32
18:30	18:45	33	-	-	-	0%	-	33
18:45	19:00	25	-	-	-	0%	-	25
19:00	19:15	8	-	-	-	0%	-	8
19:15	19:30	23	-	-	-	0%	-	23
19:30	19:45	27	-	-	-	0%	-	27
19:45	20:00	12	-	-	-	0%	-	12
Total		538	-	-	-	0%	-	538

Crossing: 2 Movement: C Date: 15/09/2015

Time		Lightweight	Trucks Trailers Buses			Heavy		Total
of	to		2 Axles	6 E	2 E	%	Total	
07:00	07:15	10	-	-	-	0%	-	10
07:15	07:30	3	-	-	-	0%	-	3
07:30	07:45	9	-	-	-	0%	-	9
07:45	08:00	10	-	-	-	0%	-	10
08:00	08:15	10	-	-	-	0%	-	10
08:15	08:30	8	-	-	-	0%	-	8
08:30	08:45	6	-	-	-	0%	-	6
08:45	09:00	14	-	-	-	0%	-	14
09:00	09:15	9	-	-	-	0%	-	9
09:15	09:30	8	-	-	-	0%	-	8
09:30	09:45	7	-	-	-	0%	-	7
09:45	10:00	9	-	-	-	0%	-	9
17:00	17:15	7	-	-	-	0%	-	7
17:15	17:30	11	-	-	-	0%	-	11
17:30	17:45	15	-	-	-	0%	-	15

17:45	18:00	13	-	-	-	0%	-	13
18:00	18:15	11	-	-	-	0%	-	11
18:15	18:30	8	-	-	-	0%	-	8
18:30	18:45	10	-	-	-	0%	-	10
18:45	19:00	9	-	-	-	0%	-	9
19:00	19:15	9	-	-	-	0%	-	9
19:15	19:30	10	-	-	-	0%	-	10
19:30	19:45	9	-	-	-	0%	-	9
19:45	20:00	2	-	-	-	0%	-	2
Total		217	-	-	-	0%	-	217

Crossing: 2 Movement: D Date: 15/09/2015

Time		Lightweight	Trucks Trailers Buses			Heavy		Total
of	to		2 Axles	6 E	2 E	%	Total	
07:00	07:15	20	5	-	1	23%	6	26
07:15	07:30	17	-	-	1	6%	1	18
07:30	07:45	12	-	-	2	14%	2	14
07:45	08:00	20	1	-	5	23%	6	26
08:00	08:15	14	2	-	2	22%	4	18
08:15	08:30	16	3	-	1	20%	4	20
08:30	08:45	28	1	-	1	7%	2	30
08:45	09:00	21	3	-	-	13%	3	24
09:00	09:15	25	2	-	3	17%	5	30
09:15	09:30	26	4	-	1	16%	5	31
09:30	09:45	21	3	-	1	16%	4	25
09:45	10:00	15	-	-	3	17%	3	18
17:00	17:15	30	-	-	1	3%	1	31
17:15	17:30	28	-	-	3	10%	3	31
17:30	17:45	25	-	-	1	4%	1	26
17:45	18:00	30	-	-	4	12%	4	34
18:00	18:15	26	-	-	1	4%	1	27
18:15	18:30	33	3	-	4	18%	7	40
18:30	18:45	53	2	-	2	7%	4	57
18:45	19:00	32	2	-	1	9%	3	35
19:00	19:15	26	-	-	2	7%	2	28
19:15	19:30	28	-	1	1	7%	2	30
19:30	19:45	32	-	-	2	6%	2	34
19:45	20:00	26	-	-	-	0%	-	26
Total		604	-	1	43	7%	44	648

Crossing: 2 Movement: E Date: 15/09/2015

Time		Lightweight	Trucks Trailers Buses			Heavy		Total
of	to		2 Axles	6 E	2 E	%	Total	
07:00	07:15	24	7	-	8	38%	15	39
07:15	07:30	32	7	-	3	24%	10	42
07:30	07:45	43	10	2	4	27%	16	59
07:45	08:00	30	7	-	5	29%	12	42
08:00	08:15	33	8	-	3	25%	11	44
08:15	08:30	20	10	2	3	43%	15	35
08:30	08:45	31	10	2	1	30%	13	44
08:45	09:00	28	8	1	7	36%	16	44
09:00	09:15	41	5	1	3	18%	9	50
09:15	09:30	45	5	-	2	13%	7	52
09:30	09:45	24	11	2	2	38%	15	39
09:45	10:00	25	8	1	4	34%	13	38
17:00	17:15	25	7	1	7	38%	15	40
17:15	17:30	30	7	2	7	35%	16	46
17:30	17:45	37	8	-	6	27%	14	51

17:45	18:00	32	5	3	3	26%	11	43
18:00	18:15	41	5	1	4	20%	10	51
18:15	18:30	34	8	-	4	26%	12	46
18:30	18:45	44	5	1	4	19%	10	54
18:45	19:00	49	9	1	2	20%	12	61
19:00	19:15	45	6	1	6	22%	13	58
19:15	19:30	33	5	1	7	28%	13	46
19:30	19:45	31	6	-	2	21%	8	39
19:45	20:00	33	2	1	3	15%	6	39
Total		810	-	23	100	13%	123	933

Crossing: 2 Movement: F Date: 15/09/2015

Time		Lightweight	Trucks Trailers Buses			Heavy		Total
of	to		2 Axles	6 E	2 E	%	Total	
07:00	07:15	5	-	-	-	0%	-	5
07:15	07:30	3	1	-	-	25%	1	4
07:30	07:45	5	-	-	-	0%	-	5
07:45	08:00	17	-	-	-	0%	-	17
08:00	08:15	13	1	-	-	7%	1	14
08:15	08:30	14	-	-	-	0%	-	14
08:30	08:45	9	-	-	-	0%	-	9
08:45	09:00	10	-	-	-	0%	-	10
09:00	09:15	16	1	-	-	6%	1	17
09:15	09:30	6	-	-	-	0%	-	6
09:30	09:45	4	2	-	-	33%	2	6
09:45	10:00	13	-	-	-	0%	-	13
17:00	17:15	16	-	-	1	6%	1	17
17:15	17:30	12	-	-	-	0%	-	12
17:30	17:45	13	-	-	1	7%	1	14
17:45	18:00	16	1	-	1	11%	2	18
18:00	18:15	16	-	-	-	0%	-	16
18:15	18:30	12	-	-	-	0%	-	12
18:30	18:45	5	-	-	-	0%	-	5
18:45	19:00	11	-	-	-	0%	-	11
19:00	19:15	16	-	-	-	0%	-	16
19:15	19:30	10	-	-	-	0%	-	10
19:30	19:45	15	-	-	-	0%	-	15
19:45	20:00	8	-	-	-	0%	-	8
Total		265	-	-	3	1%	3	268

Crossing: 2 Movement: G Date: 15/09/2015

Time		Lightweight	Trucks Trailers Buses			Heavy		Total
of	to		2 Axles	6 E	2 E	%	Total	
07:00	07:15	47	2	-	2	8%	4	51
07:15	07:30	42	1	-	-	2%	1	43
07:30	07:45	55	5	-	2	11%	7	62
07:45	08:00	44	2	-	-	4%	2	46
08:00	08:15	54	4	-	1	8%	5	59
08:15	08:30	57	1	-	1	3%	2	59
08:30	08:45	59	-	-	-	0%	-	59
08:45	09:00	61	-	-	1	2%	1	62
09:00	09:15	41	2	-	-	5%	2	43
09:15	09:30	45	2	-	1	6%	3	48
09:30	09:45	40	1	-	-	2%	1	41
09:45	10:00	59	2	-	1	5%	3	62
17:00	17:15	63	-	-	-	0%	-	63
17:15	17:30	55	-	-	2	4%	2	57
17:30	17:45	62	4	-	-	6%	4	66

17:45	18:00	59	-	-	1	2%	1	60
18:00	18:15	62	1	-	-	2%	1	63
18:15	18:30	50	-	-	1	2%	1	51
18:30	18:45	50	1	1	3	9%	5	55
18:45	19:00	60	1	-	1	3%	2	62
19:00	19:15	47	-	-	1	2%	1	48
19:15	19:30	45	-	-	1	2%	1	46
19:30	19:45	34	1	-	1	6%	2	36
19:45	20:00	27	1	-	-	4%	1	28
Total		1.218	-	1	20	2%	21	1.239

Crossing: 2 Movement: H Date: 15/09/2015

Time		Lightweight	Trucks Trailers Buses			Heavy		Total
of	to		2 Axles	6 E	2 E	%	Total	
07:00	07:15	16	-	-	1	6%	1	17
07:15	07:30	17	1	-	-	6%	1	18
07:30	07:45	16	-	-	-	0%	-	16
07:45	08:00	13	-	-	-	0%	-	13
08:00	08:15	22	-	-	2	8%	2	24
08:15	08:30	22	-	-	1	4%	1	23
08:30	08:45	19	1	-	-	5%	1	20
08:45	09:00	25	-	-	-	0%	-	25
09:00	09:15	28	1	-	1	7%	2	30
09:15	09:30	20	2	-	-	9%	2	22
09:30	09:45	15	1	-	-	6%	1	16
09:45	10:00	22	4	-	-	15%	4	26
17:00	17:15	26	1	-	1	7%	2	28
17:15	17:30	19	3	-	1	17%	4	23
17:30	17:45	29	-	-	2	6%	2	31
17:45	18:00	28	1	-	-	3%	1	29
18:00	18:15	29	-	-	1	3%	1	30
18:15	18:30	22	-	-	-	0%	-	22
18:30	18:45	28	-	-	1	3%	1	29
18:45	19:00	32	-	-	1	3%	1	33
19:00	19:15	37	-	1	-	3%	1	38
19:15	19:30	31	-	-	1	3%	1	32
19:30	19:45	23	-	-	-	0%	-	23
19:45	20:00	14	-	-	1	7%	1	15
Total		553	-	1	14	3%	15	568

Crossing: 2 Movement: I Date: 15/09/2015

Time		Lightweight	Trucks Trailers Buses			Heavy		Total
of	to		2 Axles	6 E	2 E	%	Total	
07:00	07:15	30	2	-	6	21%	8	38
07:15	07:30	20	4	-	3	26%	7	27
07:30	07:45	29	3	-	4	19%	7	36
07:45	08:00	38	2	-	2	10%	4	42
08:00	08:15	32	1	-	7	20%	8	40
08:15	08:30	27	2	-	3	16%	5	32
08:30	08:45	35	1	-	2	8%	3	38
08:45	09:00	31	1	-	7	21%	8	39
09:00	09:15	32	-	-	-	0%	-	32
09:15	09:30	22	2	-	5	24%	7	29
09:30	09:45	24	2	-	2	14%	4	28
09:45	10:00	18	3	-	4	28%	7	25
17:00	17:15	33	4	-	3	18%	7	40
17:15	17:30	34	1	-	5	15%	6	40
17:30	17:45	30	-	-	5	14%	5	35

17:45	18:00	17	1	-	6	29%	7	24
18:00	18:15	26	2	-	5	21%	7	33
18:15	18:30	27	-	-	3	10%	3	30
18:30	18:45	25	2	-	4	19%	6	31
18:45	19:00	39	-	-	8	17%	8	47
19:00	19:15	31	1	-	3	11%	4	35
19:15	19:30	30	1	-	3	12%	4	34
19:30	19:45	20	-	-	3	13%	3	23
19:45	20:00	18	-	-	4	18%	4	22
Total		668	-	-	97	13%	97	765

Crossing: 2 Movement: J Date: 15/09/2015

Time		Lightweight	Trucks Trailers Buses			Heavy		Total
of	to		2 Axles	6 E	2 E	%	Total	
07:00	07:15	2	-	-	-	0%	-	2
07:15	07:30	4	-	-	-	0%	-	4
07:30	07:45	3	-	-	-	0%	-	3
07:45	08:00	3	-	-	-	0%	-	3
08:00	08:15	3	-	-	-	0%	-	3
08:15	08:30	12	-	-	-	0%	-	12
08:30	08:45	1	-	-	-	0%	-	1
08:45	09:00	3	-	-	-	0%	-	3
09:00	09:15	3	-	-	-	0%	-	3
09:15	09:30	2	-	-	-	0%	-	2
09:30	09:45	7	-	-	-	0%	-	7
09:45	10:00	3	-	-	-	0%	-	3
17:00	17:15	4	-	-	-	0%	-	4
17:15	17:30	7	-	-	-	0%	-	7
17:30	17:45	4	-	-	-	0%	-	4
17:45	18:00	1	-	-	-	0%	-	1
18:00	18:15	3	-	-	-	0%	-	3
18:15	18:30	7	-	-	-	0%	-	7
18:30	18:45	7	-	-	-	0%	-	7
18:45	19:00	3	-	-	-	0%	-	3
19:00	19:15	7	-	-	-	0%	-	7
19:15	19:30	4	-	-	-	0%	-	4
19:30	19:45	5	-	-	-	0%	-	5
19:45	20:00	4	-	-	-	0%	-	4
Total		102	-	-	-	0%	-	102

Crossing: 3 Movement: A Date: 24/09/2015

Time		Lightweight	Trucks Trailers Buses			Heavy		Total
of	to		2 Axles	6 E	2 E	%	Total	
07:00	07:15	37	2	-	4	14%	6	43
07:15	07:30	30	4	-	-	12%	4	34
07:30	07:45	35	4	-	3	17%	7	42
07:45	08:00	39	2	-	2	9%	4	43
08:00	08:15	37	3	-	-	8%	3	40
08:15	08:30	46	4	-	-	8%	4	50
08:30	08:45	40	2	-	1	7%	3	43
08:45	09:00	46	5	-	2	13%	7	53
09:00	09:15	28	2	-	1	10%	3	31
09:15	09:30	35	1	-	1	5%	2	37
09:30	09:45	37	3	-	1	10%	4	41
09:45	10:00	43	2	-	1	7%	3	46
17:00	17:15	63	3	-	1	6%	4	67
17:15	17:30	50	1	-	3	7%	4	54
17:30	17:45	54	2	-	2	7%	4	58

17:45	18:00	60	-	-	2	3%	2	62
18:00	18:15	62	1	-	2	5%	3	65
18:15	18:30	47	2	-	2	8%	4	51
18:30	18:45	62	1	-	2	5%	3	65
18:45	19:00	62	-	-	-	0%	-	62
19:00	19:15	61	-	-	2	3%	2	63
19:15	19:30	42	-	-	1	2%	1	43
19:30	19:45	28	1	-	1	7%	2	30
19:45	20:00	27	1	-	1	7%	2	29
Total		1.071	-	-	35	3%	35	1.106

Crossing: 3 Movement: B Date: 24/09/2015

Time		Lightweight	Trucks Trailers Buses			Heavy		Total
of	to		2 Axles	6 E	2 E	%	Total	
07:00	07:15	16	2	-	-	11%	2	18
07:15	07:30	12	1	-	-	8%	1	13
07:30	07:45	9	-	-	1	10%	1	10
07:45	08:00	20	3	-	1	17%	4	24
08:00	08:15	13	-	-	-	0%	-	13
08:15	08:30	14	4	-	-	22%	4	18
08:30	08:45	18	4	-	-	18%	4	22
08:45	09:00	10	-	-	-	0%	-	10
09:00	09:15	16	1	-	1	11%	2	18
09:15	09:30	12	-	-	-	0%	-	12
09:30	09:45	8	-	-	-	0%	-	8
09:45	10:00	16	3	-	-	16%	3	19
17:00	17:15	21	-	-	-	0%	-	21
17:15	17:30	21	2	-	-	9%	2	23
17:30	17:45	18	-	-	-	0%	-	18
17:45	18:00	11	-	-	1	8%	1	12
18:00	18:15	19	-	-	-	0%	-	19
18:15	18:30	31	-	-	-	0%	-	31
18:30	18:45	17	-	-	-	0%	-	17
18:45	19:00	27	-	-	-	0%	-	27
19:00	19:15	21	1	-	-	5%	1	22
19:15	19:30	22	1	-	2	12%	3	25
19:30	19:45	21	-	-	-	0%	-	21
19:45	20:00	5	-	-	-	0%	-	5
Total		398	-	-	6	1%	6	404

Crossing: 4 Movement: A Date: 17/09/2015

Time		Lightweight	Trucks Trailers Buses			Heavy		Total
of	to		2 Axles	6 E	2 E	%	Total	
07:00	07:15	19	-	-	3	14%	3	22
07:15	07:30	15	-	-	1	6%	1	16
07:30	07:45	16	-	-	2	11%	2	18
07:45	08:00	9	1	-	4	36%	5	14
08:00	08:15	6	-	-	2	25%	2	8
08:15	08:30	14	-	-	3	18%	3	17
08:30	08:45	9	2	-	1	25%	3	12
08:45	09:00	13	2	-	4	32%	6	19
09:00	09:15	11	-	-	2	15%	2	13
09:15	09:30	9	-	-	1	10%	1	10
09:30	09:45	16	2	-	1	16%	3	19
09:45	10:00	13	-	-	5	28%	5	18
17:00	17:15	12	2	-	2	25%	4	16
17:15	17:30	16	-	-	4	20%	4	20
17:30	17:45	17	-	-	2	11%	2	19

17:45	18:00	19	-	-	3	14%	3	22
18:00	18:15	21	-	-	3	13%	3	24
18:15	18:30	17	-	-	3	15%	3	20
18:30	18:45	13	1	-	5	32%	6	19
18:45	19:00	15	1	-	2	17%	3	18
19:00	19:15	10	-	-	1	9%	1	11
19:15	19:30	5	-	-	3	38%	3	8
19:30	19:45	11	1	-	2	21%	3	14
19:45	20:00	13	-	-	4	24%	4	17
Total		319	-	-	63	16%	63	382

Crossing: 4 Movement: B Date: 17/09/2015

Time		Lightweight	Trucks Trailers Buses			Heavy		Total
of	to		2 Axles	6 E	2 E	%	Total	
07:00	07:15	20	3	-	-	13%	3	23
07:15	07:30	8	1	-	-	11%	1	9
07:30	07:45	14	2	-	-	13%	2	16
07:45	08:00	19	1	-	1	10%	2	21
08:00	08:15	12	-	-	-	0%	-	12
08:15	08:30	16	-	-	-	0%	-	16
08:30	08:45	19	1	-	-	5%	1	20
08:45	09:00	24	-	-	-	0%	-	24
09:00	09:15	12	-	-	-	0%	-	12
09:15	09:30	15	-	-	-	0%	-	15
09:30	09:45	10	1	-	-	9%	1	11
09:45	10:00	21	-	-	-	0%	-	21
17:00	17:15	26	1	-	-	4%	1	27
17:15	17:30	31	1	-	-	3%	1	32
17:30	17:45	35	-	-	1	3%	1	36
17:45	18:00	28	-	-	-	0%	-	28
18:00	18:15	36	-	-	-	0%	-	36
18:15	18:30	25	-	-	-	0%	-	25
18:30	18:45	36	-	-	-	0%	-	36
18:45	19:00	34	-	-	-	0%	-	34
19:00	19:15	28	-	-	-	0%	-	28
19:15	19:30	31	-	-	-	0%	-	31
19:30	19:45	17	-	-	-	0%	-	17
19:45	20:00	20	1	-	-	5%	1	21
Total		537	-	-	2	0%	2	539

Crossing: 4 Movement: C Date: 17/09/2015

Time		Lightweight	Trucks Trailers Buses			Heavy		Total
of	to		2 Axles	6 E	2 E	%	Total	
07:00	07:15	4	1	-	-	20%	1	5
07:15	07:30	8	1	-	-	11%	1	9
07:30	07:45	6	-	-	-	0%	-	6
07:45	08:00	5	1	-	-	17%	1	6
08:00	08:15	6	-	-	-	0%	-	6
08:15	08:30	2	-	-	-	0%	-	2
08:30	08:45	10	2	-	-	17%	2	12
08:45	09:00	4	-	-	-	0%	-	4
09:00	09:15	5	-	-	-	0%	-	5
09:15	09:30	2	-	-	-	0%	-	2
09:30	09:45	5	-	-	-	0%	-	5
09:45	10:00	4	1	-	-	20%	1	5
17:00	17:15	14	-	-	-	0%	-	14
17:15	17:30	8	-	-	-	0%	-	8
17:30	17:45	12	-	-	-	0%	-	12

17:45	18:00	10	-	-	-	0%	-	10
18:00	18:15	9	-	-	-	0%	-	9
18:15	18:30	10	-	-	-	0%	-	10
18:30	18:45	6	-	-	1	14%	1	7
18:45	19:00	6	-	-	-	0%	-	6
19:00	19:15	2	-	-	-	0%	-	2
19:15	19:30	6	-	-	-	0%	-	6
19:30	19:45	8	-	-	-	0%	-	8
19:45	20:00	5	-	-	-	0%	-	5
Total		157	-	-	1	1%	1	158

Crossing: 4 Movement: D Date: 17/09/2015

Time		Lightweight	Trucks Trailers Buses			Heavy		Total
of	to		2 Axles	6 E	2 E	%	Total	
07:00	07:15	-	-	-	-	0%	-	-
07:15	07:30	-	-	-	-	0%	-	-
07:30	07:45	-	-	-	-	0%	-	-
07:45	08:00	-	-	-	-	0%	-	-
08:00	08:15	1	-	-	-	0%	-	1
08:15	08:30	-	-	-	-	0%	-	-
08:30	08:45	1	-	-	-	0%	-	1
08:45	09:00	1	-	-	-	0%	-	1
09:00	09:15	1	-	-	-	0%	-	1
09:15	09:30	-	-	-	-	0%	-	-
09:30	09:45	-	-	-	-	0%	-	-
09:45	10:00	4	-	-	-	0%	-	4
17:00	17:15	-	-	-	-	0%	-	-
17:15	17:30	-	-	-	-	0%	-	-
17:30	17:45	-	-	-	-	0%	-	-
17:45	18:00	2	-	-	-	0%	-	2
18:00	18:15	1	-	-	-	0%	-	1
18:15	18:30	1	-	-	-	0%	-	1
18:30	18:45	-	-	-	-	0%	-	-
18:45	19:00	-	-	-	1	100%	1	1
19:00	19:15	-	-	-	-	0%	-	-
19:15	19:30	1	-	-	-	0%	-	1
19:30	19:45	-	-	-	-	0%	-	-
19:45	20:00	-	-	-	-	0%	-	-
Total		13	-	-	1	7%	1	14

Crossing: 4 Movement: E Date: 17/09/2015

Time		Lightweight	Trucks Trailers Buses			Heavy		Total
of	to		2 Axles	6 E	2 E	%	Total	
07:00	07:15	6	1	-	-	0%	1	7
07:15	07:30	6	1	-	-	0%	1	7
07:30	07:45	10	2	-	-	0%	2	12
07:45	08:00	12	-	-	-	0%	-	12
08:00	08:15	12	1	-	1	0%	2	14
08:15	08:30	5	1	-	-	0%	1	6
08:30	08:45	10	-	-	-	0%	-	10
08:45	09:00	13	2	-	-	0%	2	15
09:00	09:15	4	-	-	-	0%	-	4
09:15	09:30	4	3	-	-	0%	3	7
09:30	09:45	11	-	-	-	0%	-	11
09:45	10:00	14	3	-	-	0%	3	17
17:00	17:15	7	-	-	-	0%	-	7
17:15	17:30	10	-	-	-	0%	-	10
17:30	17:45	15	-	-	-	0%	-	15

17:45	18:00	11	-	-	-	0%	-	11
18:00	18:15	12	-	-	-	0%	-	12
18:15	18:30	10	-	-	-	0%	-	10
18:30	18:45	12	-	-	-	0%	-	12
18:45	19:00	10	-	-	-	100%	-	10
19:00	19:15	18	-	-	-	0%	-	18
19:15	19:30	11	-	-	-	0%	-	11
19:30	19:45	13	-	-	-	0%	-	13
19:45	20:00	11	-	-	-	0%	-	11
Total		247	-	-	1	0%	1	248

Crossing: 4 Movement: F Date: 17/09/2015

Time		Lightweight	Trucks Trailers Buses			Heavy		Total
of	to		2 Axles	6 E	2 E	%	Total	
07:00	07:15	9	1	-	1	18%	2	11
07:15	07:30	9	1	-	2	25%	3	12
07:30	07:45	5	-	-	1	17%	1	6
07:45	08:00	9	2	-	-	18%	2	11
08:00	08:15	7	3	-	-	30%	3	10
08:15	08:30	2	2	-	3	71%	5	7
08:30	08:45	14	-	-	-	0%	-	14
08:45	09:00	10	1	-	-	9%	1	11
09:00	09:15	9	2	-	1	25%	3	12
09:15	09:30	11	2	-	-	15%	2	13
09:30	09:45	4	1	-	1	33%	2	6
09:45	10:00	14	3	-	1	22%	4	18
17:00	17:15	20	-	-	1	0%	1	21
17:15	17:30	7	1	-	5	46%	6	13
17:30	17:45	13	-	-	2	13%	2	15
17:45	18:00	15	-	-	2	12%	2	17
18:00	18:15	17	-	-	2	11%	2	19
18:15	18:30	12	1	-	-	8%	1	13
18:30	18:45	18	-	-	2	10%	2	20
18:45	19:00	11	-	-	-	0%	-	11
19:00	19:15	18	-	-	2	10%	2	20
19:15	19:30	20	-	-	-	0%	-	20
19:30	19:45	13	-	-	1	7%	1	14
19:45	20:00	9	-	-	1	10%	1	10
Total		276	-	-	28	9%	28	304

Crossing: 4 Movement: G Date: 17/09/2015

Time		Lightweight	Trucks Trailers Buses			Heavy		Total
of	to		2 Axles	6 E	2 E	%	Total	
07:00	07:15	11	-	-	-	0%	-	11
07:15	07:30	6	1	-	-	14%	1	7
07:30	07:45	9	1	-	-	10%	1	10
07:45	08:00	5	-	-	-	0%	-	5
08:00	08:15	12	-	-	-	0%	-	12
08:15	08:30	5	-	-	-	0%	-	5
08:30	08:45	22	1	-	1	8%	2	24
08:45	09:00	9	-	-	-	0%	-	9
09:00	09:15	7	-	-	-	0%	-	7
09:15	09:30	8	1	-	-	11%	1	9
09:30	09:45	4	-	-	-	0%	-	4
09:45	10:00	4	-	-	-	0%	-	4
17:00	17:15	8	-	-	-	0%	-	8
17:15	17:30	12	1	-	-	8%	1	13
17:30	17:45	10	-	-	-	0%	-	10

17:45	18:00	10	2	-	-	17%	2	12
18:00	18:15	11	1	-	-	8%	1	12
18:15	18:30	7	-	-	-	0%	-	7
18:30	18:45	18	-	-	-	0%	-	18
18:45	19:00	21	-	-	-	0%	-	21
19:00	19:15	12	-	-	-	0%	-	12
19:15	19:30	14	-	-	-	0%	-	14
19:30	19:45	10	-	-	-	0%	-	10
19:45	20:00	10	-	-	-	0%	-	10
Total		245	-	-	1	0%	1	246

Crossing:	4	Movement:	H			Date:	17/09/2015	
Time		Lightweight	Trucks Trailers Buses			Heavy		Total
of	to		2 Axles	6 E	2 E	%	Total	
07:00	07:15	1	-	-	-	0%	-	1
07:15	07:30	1	1	-	-	50%	1	2
07:30	07:45	2	-	-	-	0%	-	2
07:45	08:00	3	-	-	1	25%	1	4
08:00	08:15	6	-	-	-	0%	-	6
08:15	08:30	4	1	-	-	20%	1	5
08:30	08:45	3	1	-	-	25%	1	4
08:45	09:00	4	1	-	-	20%	1	5
09:00	09:15	2	-	-	-	0%	-	2
09:15	09:30	2	1	-	-	33%	1	3
09:30	09:45	1	-	-	-	0%	-	1
09:45	10:00	2	-	-	-	0%	-	2
17:00	17:15	-	-	-	-	0%	-	-
17:15	17:30	2	-	-	1	33%	1	3
17:30	17:45	3	-	-	-	0%	-	3
17:45	18:00	2	-	1	-	33%	1	3
18:00	18:15	2	-	-	-	0%	-	2
18:15	18:30	3	-	-	2	40%	2	5
18:30	18:45	5	-	-	-	0%	-	5
18:45	19:00	2	-	-	-	0%	-	2
19:00	19:15	1	-	-	1	50%	1	2
19:15	19:30	1	-	-	-	0%	-	1
19:30	19:45	1	-	-	-	0%	-	1
19:45	20:00	4	-	-	-	0%	-	4
Total		57	-	1	5	10%	6	63

Crossing:	4	Movement:	I			Date:	17/09/2015	
Time		Lightweight	Trucks Trailers Buses			Heavy		Total
of	to		2 Axles	6 E	2 E	%	Total	
07:00	07:15	6	-	-	-	0%	-	6
07:15	07:30	4	1	-	-	20%	1	5
07:30	07:45	4	-	-	-	0%	-	4
07:45	08:00	1	-	-	-	0%	-	1
08:00	08:15	9	1	-	-	10%	1	10
08:15	08:30	9	-	-	1	10%	1	10
08:30	08:45	5	-	-	-	0%	-	5
08:45	09:00	3	2	-	-	40%	2	5
09:00	09:15	4	-	-	-	0%	-	4
09:15	09:30	7	1	-	-	13%	1	8
09:30	09:45	3	1	-	-	25%	1	4
09:45	10:00	4	-	-	-	0%	-	4
17:00	17:15	11	-	-	-	0%	-	11
17:15	17:30	4	1	-	-	20%	1	5
17:30	17:45	17	-	-	-	0%	-	17

17:45	18:00	11	-	-	1	8%	1	12
18:00	18:15	10	-	-	-	0%	-	10
18:15	18:30	13	-	-	-	0%	-	13
18:30	18:45	5	-	-	-	0%	-	5
18:45	19:00	7	-	-	-	0%	-	7
19:00	19:15	7	-	-	-	0%	-	7
19:15	19:30	10	-	-	-	0%	-	10
19:30	19:45	6	-	-	-	0%	-	6
19:45	20:00	6	-	-	-	0%	-	6
Total		166	-	-	2	1%	2	168

Crossing: 4 Movement: J Date: 17/09/2015

Time		Lightweight	Trucks Trailers Buses			Heavy		Total
of	to		2 Axles	6 E	2 E	%	Total	
07:00	07:15	21	-	-	-	0%	-	21
07:15	07:30	18	1	-	-	5%	1	19
07:30	07:45	36	1	-	-	3%	1	37
07:45	08:00	24	-	-	-	0%	-	24
08:00	08:15	25	1	-	-	4%	1	26
08:15	08:30	25	-	-	-	0%	-	25
08:30	08:45	43	2	-	-	4%	2	45
08:45	09:00	31	2	-	1	9%	3	34
09:00	09:15	14	1	-	-	7%	1	15
09:15	09:30	26	1	-	-	4%	1	27
09:30	09:45	26	1	-	-	4%	1	27
09:45	10:00	30	2	-	-	6%	2	32
17:00	17:15	55	2	-	-	0%	2	57
17:15	17:30	38	-	-	-	0%	-	38
17:30	17:45	70	1	-	-	1%	1	71
17:45	18:00	70	1	-	-	1%	1	71
18:00	18:15	70	-	-	-	0%	-	70
18:15	18:30	55	1	-	-	2%	1	56
18:30	18:45	68	-	-	-	0%	-	68
18:45	19:00	33	-	-	-	0%	-	33
19:00	19:15	50	1	-	-	2%	1	51
19:15	19:30	30	-	-	-	0%	-	30
19:30	19:45	26	-	-	-	0%	-	26
19:45	20:00	21	-	-	-	0%	-	21
Total		905	-	-	1	0%	1	906

Crossing: 4 Movement: K Date: 17/09/2015

Time		Lightweight	Trucks Trailers Buses			Heavy		Total
of	to		2 Axles	6 E	2 E	%	Total	
07:00	07:15	15	-	-	-	0%	-	15
07:15	07:30	16	-	-	-	0%	-	16
07:30	07:45	16	1	-	2	16%	3	19
07:45	08:00	13	-	-	1	7%	1	14
08:00	08:15	18	-	-	-	0%	-	18
08:15	08:30	16	1	-	-	6%	1	17
08:30	08:45	15	2	-	1	17%	3	18
08:45	09:00	9	-	-	-	0%	-	9
09:00	09:15	18	1	-	-	5%	1	19
09:15	09:30	13	-	-	1	7%	1	14
09:30	09:45	8	4	-	-	33%	4	12
09:45	10:00	14	-	-	-	0%	-	14
17:00	17:15	14	-	-	1	0%	1	15
17:15	17:30	15	2	-	2	21%	4	19
17:30	17:45	16	-	-	-	0%	-	16

17:45	18:00	21	-	-	1	5%	1	22
18:00	18:15	25	-	-	-	0%	-	25
18:15	18:30	12	-	-	-	0%	-	12
18:30	18:45	21	-	-	-	0%	-	21
18:45	19:00	15	-	-	-	0%	-	15
19:00	19:15	26	1	-	1	7%	2	28
19:15	19:30	16	-	-	-	0%	-	16
19:30	19:45	8	-	-	-	0%	-	8
19:45	20:00	16	1	-	1	11%	2	18
Total		376	-	-	11	3%	11	387

Crossing: 4 Motion: L Date: 10/09/2015

Time		Lightweight	Trucks Trailers Buses			Heavy		Total
of	to		2 Axles	6 E	2 E	%	Total	
07:00	07:15	-	-	-	-	0%	-	-
07:15	07:30	-	-	-	-	0%	-	-
07:30	07:45	-	-	-	-	0%	-	-
07:45	08:00	-	-	-	-	0%	-	-
08:00	08:15	1	-	-	-	0%	-	1
08:15	08:30	1	-	-	-	0%	-	1
08:30	08:45	1	-	-	-	0%	-	1
08:45	09:00	2	-	-	-	0%	-	2
09:00	09:15	-	-	-	-	0%	-	-
09:15	09:30	-	-	-	-	0%	-	-
09:30	09:45	-	-	-	-	0%	-	-
09:45	10:00	-	-	-	-	0%	-	-
17:00	17:15	1	-	-	-	0%	-	1
17:15	17:30	1	1	-	-	50%	1	2
17:30	17:45	1	-	-	-	0%	-	1
17:45	18:00	-	-	-	-	0%	-	-
18:00	18:15	3	-	-	-	0%	-	3
18:15	18:30	2	-	-	-	0%	-	2
18:30	18:45	-	-	-	-	0%	-	-
18:45	19:00	3	-	-	-	0%	-	3
19:00	19:15	2	-	-	-	0%	-	2
19:15	19:30	2	-	-	-	0%	-	2
19:30	19:45	2	-	-	-	0%	-	2
19:45	20:00	1	-	-	-	0%	-	1
Total		23	1	-	-	4%	1	24

Crossing: 4 Movement: M Date: 17/09/2015

Time		Lightweight	Trucks Trailers Buses			Heavy		Total
of	to		2 Axles	6 E	2 E	%	Total	
07:00	07:15	5	-	-	-	0%	-	5
07:15	07:30	6	-	-	-	0%	-	6
07:30	07:45	7	-	-	-	0%	-	7
07:45	08:00	11	-	-	-	0%	-	11
08:00	08:15	5	1	-	-	17%	1	6
08:15	08:30	4	-	-	-	0%	-	4
08:30	08:45	10	-	-	-	0%	-	10
08:45	09:00	6	-	-	-	0%	-	6
09:00	09:15	3	-	-	-	0%	-	3
09:15	09:30	1	1	-	-	50%	1	2
09:30	09:45	3	2	-	-	40%	2	5
09:45	10:00	4	1	-	-	20%	1	5

17:00	17:15	11	-	-	-	0%	-	11
17:15	17:30	3	-	-	-	0%	-	3
17:30	17:45	9	-	-	-	0%	-	9
17:45	18:00	10	-	-	-	0%	-	10
18:00	18:15	7	-	-	-	0%	-	7
18:15	18:30	7	-	-	-	0%	-	7
18:30	18:45	7	-	-	-	0%	-	7
18:45	19:00	10	-	-	-	0%	-	10
19:00	19:15	2	1	-	-	33%	1	3
19:15	19:30	3	-	-	-	0%	-	3
19:30	19:45	8	-	-	-	0%	-	8
19:45	20:00	10	-	-	-	0%	-	10
Total		152	-	-	-	4%	-	152

Crossing: 4 Movement: N Date: 17/09/2015

Time		Lightweight	Trucks Trailers Buses			Heavy		Total
of	to		2 Axles	6 E	2 E	%	Total	
07:00	07:15	17	-	-	-	0%	-	17
07:15	07:30	23	1	-	2	12%	3	26
07:30	07:45	25	4	-	1	17%	5	30
07:45	08:00	6	3	-	-	33%	3	9
08:00	08:15	14	1	-	1	13%	2	16
08:15	08:30	11	2	-	-	15%	2	13
08:30	08:45	18	-	-	1	5%	1	19
08:45	09:00	15	2	-	2	21%	4	19
09:00	09:15	10	3	-	1	29%	4	14
09:15	09:30	11	1	-	-	8%	1	12
09:30	09:45	17	-	-	1	6%	1	18
09:45	10:00	17	1	-	1	11%	2	19
17:00	17:15	9	1	-	1	18%	2	11
17:15	17:30	16	1	-	2	16%	3	19
17:30	17:45	9	-	-	-	0%	-	9
17:45	18:00	17	-	-	1	6%	1	18
18:00	18:15	24	-	-	2	8%	2	26
18:15	18:30	9	-	-	-	0%	-	9
18:30	18:45	32	-	-	1	3%	1	33
18:45	19:00	14	1	-	1	13%	2	16
19:00	19:15	23	-	-	-	0%	-	23
19:15	19:30	13	1	-	1	13%	2	15
19:30	19:45	19	-	-	-	0%	-	19
19:45	20:00	18	-	-	-	0%	-	18
Total		387	-	-	19	4%	19	406

Crossing: 4 Movement: O Date: 17/09/2015

Time		Lightweight	Trucks Trailers Buses			Heavy		Total
of	to		2 Axles	6 E	2 E	%	Total	
07:00	07:15	3	1	-	-	25%	1	4
07:15	07:30	2	-	-	-	0%	-	2
07:30	07:45	3	-	-	-	0%	-	3
07:45	08:00	2	-	-	-	0%	-	2
08:00	08:15	4	-	-	-	0%	-	4
08:15	08:30	3	-	-	-	0%	-	3
08:30	08:45	9	-	-	-	0%	-	9
08:45	09:00	8	1	-	-	11%	1	9
09:00	09:15	4	-	-	-	0%	-	4
09:15	09:30	2	1	-	-	33%	1	3
09:30	09:45	1	-	-	-	0%	-	1
09:45	10:00	5	-	-	-	0%	-	5

17:00	17:15	2	-	-	-	0%	-	2
17:15	17:30	7	-	-	-	0%	-	7
17:30	17:45	6	-	-	-	0%	-	6
17:45	18:00	4	-	-	-	0%	-	4
18:00	18:15	3	-	-	-	0%	-	3
18:15	18:30	4	-	-	-	0%	-	4
18:30	18:45	6	-	-	-	0%	-	6
18:45	19:00	2	-	-	-	0%	-	2
19:00	19:15	2	-	-	-	0%	-	2
19:15	19:30	6	-	-	-	0%	-	6
19:30	19:45	3	-	-	-	0%	-	3
19:45	20:00	2	-	-	-	0%	-	2
Total		93	-	-	-	4%	-	93

Crossing: 4 Movement: P Date: 17/09/2015

Time		Lightweight	Trucks Trailers Buses			Heavy		Total
of	to		2 Axles	6 E	2 E	%	Total	
07:00	07:15	-	-	-	-	0%	-	-
07:15	07:30	-	-	-	-	0%	-	-
07:30	07:45	-	-	-	-	0%	-	-
07:45	08:00	-	-	-	-	0%	-	-
08:00	08:15	-	-	-	-	0%	-	-
08:15	08:30	1	-	-	-	0%	-	1
08:30	08:45	-	-	-	-	0%	-	-
08:45	09:00	-	-	-	-	0%	-	-
09:00	09:15	-	-	-	-	0%	-	-
09:15	09:30	-	-	-	-	0%	-	-
09:30	09:45	-	-	-	-	0%	-	-
09:45	10:00	-	-	-	-	0%	-	-
17:00	17:15	1	-	-	-	0%	-	1
17:15	17:30	-	-	-	-	0%	-	-
17:30	17:45	-	-	-	-	0%	-	-
17:45	18:00	1	-	-	-	0%	-	1
18:00	18:15	1	-	-	-	0%	-	1
18:15	18:30	-	-	-	-	0%	-	-
18:30	18:45	-	-	-	-	0%	-	-
18:45	19:00	1	-	-	-	0%	-	1
19:00	19:15	-	-	-	-	0%	-	-
19:15	19:30	-	-	-	-	0%	-	-
19:30	19:45	-	-	-	-	0%	-	-
19:45	20:00	-	-	-	-	0%	-	-
Total		5	-	-	-	0%	-	5

Crossing: 5 Movement: A Date: 15/09/2015

Time		Lightweight	Trucks Trailers Buses			Heavy		Total
of	to		2 Axles	6 E	2 E	%	Total	
07:00	07:15	3	-	-	-	0%	-	3
07:15	07:30	7	-	-	-	0%	-	7
07:30	07:45	12	-	-	-	0%	-	12
07:45	08:00	4	-	-	-	0%	-	4
08:00	08:15	4	1	-	-	20%	1	5
08:15	08:30	8	-	-	-	0%	-	8
08:30	08:45	5	-	-	-	0%	-	5
08:45	09:00	4	-	-	-	0%	-	4
09:00	09:15	3	1	-	-	25%	1	4
09:15	09:30	4	-	-	-	0%	-	4
09:30	09:45	8	-	-	-	0%	-	8
09:45	10:00	-	1	-	-	100%	1	1

of	to	Lightweight	2 Axles	6 E	2 E	%	Total	Total
17:00	17:15	11	-	-	-	0%	-	11
17:15	17:30	15	1	-	-	6%	1	16
17:30	17:45	10	-	-	-	0%	-	10
17:45	18:00	3	-	-	-	0%	-	3
18:00	18:15	8	-	-	-	0%	-	8
18:15	18:30	11	-	-	-	0%	-	11
18:30	18:45	6	-	-	-	0%	-	6
18:45	19:00	7	-	-	-	0%	-	7
19:00	19:15	11	-	-	-	0%	-	11
19:15	19:30	3	-	-	-	0%	-	3
19:30	19:45	4	-	-	-	0%	-	4
19:45	20:00	3	-	-	-	0%	-	3
Total		154	-	-	-	0%	-	154

Crossing: 5 Movement: B Date: 15/09/2015

Time		Lightweight	Trucks Trailers Buses			Heavy		Total
of	to		2 Axles	6 E	2 E	%	Total	
07:00	07:15	21	-	-	-	0%	-	21
07:15	07:30	15	-	-	-	0%	-	15
07:30	07:45	24	-	-	-	0%	-	24
07:45	08:00	11	-	-	-	0%	-	11
08:00	08:15	16	1	-	-	6%	1	17
08:15	08:30	13	-	-	-	0%	-	13
08:30	08:45	21	1	-	-	5%	1	22
08:45	09:00	11	1	-	-	8%	1	12
09:00	09:15	10	-	-	-	0%	-	10
09:15	09:30	12	-	-	-	0%	-	12
09:30	09:45	7	1	-	-	13%	1	8
09:45	10:00	11	-	-	-	0%	-	11
17:00	17:15	30	-	-	-	0%	-	30
17:15	17:30	21	-	-	-	0%	-	21
17:30	17:45	16	-	-	-	0%	-	16
17:45	18:00	16	-	-	-	0%	-	16
18:00	18:15	17	-	-	-	0%	-	17
18:15	18:30	11	-	-	-	0%	-	11
18:30	18:45	11	1	-	-	8%	1	12
18:45	19:00	24	-	-	-	0%	-	24
19:00	19:15	17	-	-	-	0%	-	17
19:15	19:30	9	-	-	-	0%	-	9
19:30	19:45	10	-	-	-	0%	-	10
19:45	20:00	3	-	-	-	0%	-	3
Total		357	-	-	-	0%	-	357

Crossing: 5 Movement: C Date: 15/09/2015

Time		Lightweight	Trucks Trailers Buses			Heavy		Total
of	to		2 Axles	6 E	2 E	%	Total	
07:00	07:15	3	-	-	-	0%	-	3
07:15	07:30	1	-	-	-	0%	-	1
07:30	07:45	2	-	-	-	0%	-	2
07:45	08:00	2	-	-	-	0%	-	2
08:00	08:15	3	-	-	-	0%	-	3
08:15	08:30	5	-	-	-	0%	-	5
08:30	08:45	6	-	-	-	0%	-	6
08:45	09:00	5	-	-	-	0%	-	5
09:00	09:15	3	-	-	-	0%	-	3
09:15	09:30	5	-	-	-	0%	-	5
09:30	09:45	3	-	-	-	0%	-	3
09:45	10:00	3	-	-	-	0%	-	3

17:00	17:15	3	-	-	-	0%	-	3
17:15	17:30	3	-	-	-	0%	-	3
17:30	17:45	3	-	-	-	0%	-	3
17:45	18:00	4	-	-	-	0%	-	4
18:00	18:15	3	-	-	-	0%	-	3
18:15	18:30	5	-	-	-	0%	-	5
18:30	18:45	3	-	-	-	0%	-	3
18:45	19:00	3	-	-	-	0%	-	3
19:00	19:15	2	-	-	-	0%	-	2
19:15	19:30	3	-	-	-	0%	-	3
19:30	19:45	3	-	-	-	0%	-	3
19:45	20:00	4	-	-	-	0%	-	4
Total		80	-	-	-	0%	-	80

Crossing: 5 Movement: D Date: 15/09/2015

Time		Lightweight	Trucks Trailers Buses			Heavy		Total
of	to		2 Axles	6 E	2 E	%	Total	
07:00	07:15	-	-	-	-	0%	-	-
07:15	07:30	-	-	-	-	0%	-	-
07:30	07:45	-	-	-	-	0%	-	-
07:45	08:00	-	-	-	-	0%	-	-
08:00	08:15	-	-	-	-	0%	-	-
08:15	08:30	-	-	-	-	0%	-	-
08:30	08:45	-	-	-	-	0%	-	-
08:45	09:00	1	-	-	-	0%	-	1
09:00	09:15	-	-	-	-	0%	-	-
09:15	09:30	-	-	-	-	0%	-	-
09:30	09:45	-	-	-	-	0%	-	-
09:45	10:00	-	-	-	-	0%	-	-
17:00	17:15	1	-	-	-	0%	-	1
17:15	17:30	-	-	-	-	0%	-	-
17:30	17:45	1	-	-	-	0%	-	1
17:45	18:00	3	-	-	-	0%	-	3
18:00	18:15	-	-	-	-	0%	-	-
18:15	18:30	-	-	-	-	0%	-	-
18:30	18:45	-	-	-	-	0%	-	-
18:45	19:00	-	-	-	-	0%	-	-
19:00	19:15	-	-	-	-	0%	-	-
19:15	19:30	-	-	-	-	0%	-	-
19:30	19:45	-	-	-	-	0%	-	-
19:45	20:00	-	-	-	-	0%	-	-
Total		6	-	-	-	0%	-	6

Crossing: 5 Movement: E Date: 15/09/2015

Time		Lightweight	Trucks Trailers Buses			Heavy		Total
of	to		2 Axles	6 E	2 E	%	Total	
07:00	07:15	12	1	-	3	25%	4	16
07:15	07:30	7	-	-	2	22%	2	9
07:30	07:45	16	-	-	1	6%	1	17
07:45	08:00	8	-	-	5	38%	5	13
08:00	08:15	17	1	-	3	19%	4	21
08:15	08:30	16	-	-	1	6%	1	17
08:30	08:45	12	1	-	4	29%	5	17
08:45	09:00	10	1	-	2	23%	3	13
09:00	09:15	15	3	-	2	25%	5	20
09:15	09:30	14	4	-	2	30%	6	20
09:30	09:45	8	-	-	3	27%	3	11
09:45	10:00	12	-	-	2	14%	2	14

17:00	17:15	26	-	-	2	7%	2	28
17:15	17:30	29	1	-	3	12%	4	33
17:30	17:45	37	-	-	5	12%	5	42
17:45	18:00	20	-	-	4	17%	4	24
18:00	18:15	30	2	-	1	9%	3	33
18:15	18:30	26	-	-	5	16%	5	31
18:30	18:45	33	2	-	5	18%	7	40
18:45	19:00	24	1	-	2	11%	3	27
19:00	19:15	20	2	-	2	17%	4	24
19:15	19:30	12	-	-	4	25%	4	16
19:30	19:45	13	-	-	-	0%	-	13
19:45	20:00	16	-	-	3	16%	3	19
Total		433	-	-	66	0%	66	499

Crossing: 5 Movement: F Date: 15/09/2015

Time		Lightweight	Trucks Trailers Buses			Heavy		Total
of	to		2 Axles	6 E	2 E	%	Total	
07:00	07:15	12	-	-	-	0%	-	12
07:15	07:30	8	2	-	-	20%	2	10
07:30	07:45	11	-	-	-	0%	-	11
07:45	08:00	10	1	-	-	9%	1	11
08:00	08:15	2	-	-	-	0%	-	2
08:15	08:30	9	-	-	-	0%	-	9
08:30	08:45	10	-	-	-	0%	-	10
08:45	09:00	9	-	-	-	0%	-	9
09:00	09:15	7	1	-	-	13%	1	8
09:15	09:30	6	-	-	-	0%	-	6
09:30	09:45	12	-	-	-	0%	-	12
09:45	10:00	9	-	-	-	0%	-	9
17:00	17:15	13	-	-	1	7%	1	14
17:15	17:30	24	-	-	-	0%	-	24
17:30	17:45	25	-	-	-	0%	-	25
17:45	18:00	17	1	-	-	6%	1	18
18:00	18:15	11	-	-	-	0%	-	11
18:15	18:30	15	-	-	-	0%	-	15
18:30	18:45	12	-	-	-	0%	-	12
18:45	19:00	10	-	-	-	0%	-	10
19:00	19:15	20	-	-	-	0%	-	20
19:15	19:30	11	-	-	-	0%	-	11
19:30	19:45	7	-	-	-	0%	-	7
19:45	20:00	5	-	-	-	0%	-	5
Total		275	-	-	1	0%	1	276

Crossing: 5 Movement: G Date: 15/09/2015

Time		Lightweight	Trucks Trailers Buses			Heavy		Total
of	to		2 Axles	6 E	2 E	%	Total	
07:00	07:15	3	-	-	-	0%	-	3
07:15	07:30	7	-	-	-	0%	-	7
07:30	07:45	7	-	-	-	0%	-	7
07:45	08:00	6	-	-	-	0%	-	6
08:00	08:15	9	-	-	-	0%	-	9
08:15	08:30	6	2	-	-	25%	2	8
08:30	08:45	3	-	-	-	0%	-	3
08:45	09:00	7	-	-	-	0%	-	7
09:00	09:15	3	-	-	-	0%	-	3
09:15	09:30	9	1	-	-	10%	1	10
09:30	09:45	6	1	-	-	14%	1	7
09:45	10:00	3	-	-	-	0%	-	3

17:00	17:15	7	-	-	-	0%	-	7
17:15	17:30	18	-	-	-	0%	-	18
17:30	17:45	11	-	-	-	0%	-	11
17:45	18:00	11	1	-	-	8%	1	12
18:00	18:15	14	-	-	-	0%	-	14
18:15	18:30	8	-	-	-	0%	-	8
18:30	18:45	22	-	-	-	0%	-	22
18:45	19:00	13	-	-	-	0%	-	13
19:00	19:15	11	-	-	-	0%	-	11
19:15	19:30	9	-	-	-	0%	-	9
19:30	19:45	9	-	-	-	0%	-	9
19:45	20:00	7	-	-	-	0%	-	7
Total		209	-	-	-	0%	-	209

Crossing: 5 Movement: H Date: 15/09/2015

Time		Lightweight	Trucks Trailers Buses			Heavy		Total
of	to		2 Axles	6 E	2 E	%	Total	
07:00	07:15	-	-	-	-	0%	-	-
07:15	07:30	2	-	-	-	0%	-	2
07:30	07:45	2	-	-	-	0%	-	2
07:45	08:00	2	-	-	-	0%	-	2
08:00	08:15	2	-	-	-	0%	-	2
08:15	08:30	4	-	-	-	0%	-	4
08:30	08:45	-	-	-	-	0%	-	-
08:45	09:00	3	-	-	-	0%	-	3
09:00	09:15	3	-	-	-	0%	-	3
09:15	09:30	2	-	-	-	0%	-	2
09:30	09:45	4	-	-	-	0%	-	4
09:45	10:00	2	-	-	-	0%	-	2
17:00	17:15	3	-	-	-	0%	-	3
17:15	17:30	4	-	-	-	0%	-	4
17:30	17:45	4	-	-	-	0%	-	4
17:45	18:00	5	-	-	-	0%	-	5
18:00	18:15	6	-	-	-	0%	-	6
18:15	18:30	3	-	-	-	0%	-	3
18:30	18:45	2	-	-	-	0%	-	2
18:45	19:00	4	-	-	-	0%	-	4
19:00	19:15	2	-	-	-	0%	-	2
19:15	19:30	2	-	-	-	0%	-	2
19:30	19:45	1	-	-	-	0%	-	1
19:45	20:00	1	-	-	-	0%	-	1
Total		63	-	-	-	0%	-	63

Crossing: 5 Movement: I Date: 15/09/2015

Time		Lightweight	Trucks Trailers Buses			Heavy		Total
of	to		2 Axles	6 E	2 E	%	Total	
07:00	07:15	4	-	-	-	0%	-	4
07:15	07:30	3	-	-	-	0%	-	3
07:30	07:45	11	1	-	1	15%	2	13
07:45	08:00	8	-	-	-	0%	-	8
08:00	08:15	5	-	-	-	0%	-	5
08:15	08:30	6	-	-	-	0%	-	6
08:30	08:45	8	-	-	-	0%	-	8
08:45	09:00	6	-	-	-	0%	-	6
09:00	09:15	8	-	-	-	0%	-	8
09:15	09:30	5	-	-	-	0%	-	5
09:30	09:45	5	-	-	1	17%	1	6
09:45	10:00	7	-	-	-	0%	-	7

17:00	17:15	11	-	-	-	0%	-	11
17:15	17:30	8	-	-	-	0%	-	8
17:30	17:45	12	-	-	-	0%	-	12
17:45	18:00	9	1	-	-	10%	1	10
18:00	18:15	14	-	-	-	0%	-	14
18:15	18:30	16	-	-	-	0%	-	16
18:30	18:45	9	-	-	-	0%	-	9
18:45	19:00	10	-	-	-	0%	-	10
19:00	19:15	14	-	-	-	0%	-	14
19:15	19:30	5	-	-	-	0%	-	5
19:30	19:45	9	-	-	-	0%	-	9
19:45	20:00	5	-	-	-	0%	-	5
Total		198	-	-	2	0%	2	200

Crossing: 5 Movement: J Date: 15/09/2015

Time		Lightweight	Trucks Trailers Buses			Heavy		Total
of	to		2 Axles	6 E	2 E	%	Total	
07:00	07:15	24	1	-	-	4%	1	25
07:15	07:30	13	-	-	1	7%	1	14
07:30	07:45	20	-	-	-	0%	-	20
07:45	08:00	18	2	-	-	10%	2	20
08:00	08:15	18	-	-	-	0%	-	18
08:15	08:30	19	-	-	-	0%	-	19
08:30	08:45	19	-	-	-	0%	-	19
08:45	09:00	23	-	-	-	0%	-	23
09:00	09:15	14	2	-	-	13%	2	16
09:15	09:30	20	-	-	-	0%	-	20
09:30	09:45	14	1	-	-	7%	1	15
09:45	10:00	16	1	-	-	6%	1	17
17:00	17:15	21	1	-	-	5%	1	22
17:15	17:30	33	-	-	-	0%	-	33
17:30	17:45	35	1	-	-	3%	1	36
17:45	18:00	32	-	-	-	0%	-	32
18:00	18:15	49	-	-	-	0%	-	49
18:15	18:30	39	-	-	-	0%	-	39
18:30	18:45	36	1	-	-	3%	1	37
18:45	19:00	21	-	-	-	0%	-	21
19:00	19:15	22	-	-	-	0%	-	22
19:15	19:30	21	-	-	-	0%	-	21
19:30	19:45	21	-	-	-	0%	-	21
19:45	20:00	14	-	-	-	0%	-	14
Total		562	-	-	1	0%	1	563

Crossing: 5 Movement: K Date: 15/09/2015

Time		Lightweight	Trucks Trailers Buses			Heavy		Total
of	to		2 Axles	6 E	2 E	%	Total	
07:00	07:15	9	1	-	1	18%	2	11
07:15	07:30	9	1	-	-	10%	1	10
07:30	07:45	14	-	-	-	0%	-	14
07:45	08:00	20	-	-	-	0%	-	20
08:00	08:15	11	1	-	-	8%	1	12
08:15	08:30	8	1	-	-	11%	1	9
08:30	08:45	13	2	-	-	13%	2	15
08:45	09:00	10	3	-	-	23%	3	13
09:00	09:15	12	1	-	-	8%	1	13
09:15	09:30	12	1	-	1	14%	2	14
09:30	09:45	15	1	-	-	6%	1	16
09:45	10:00	15	-	-	-	0%	-	15

17:00	17:15	28	-	-	-	0%	-	28
17:15	17:30	29	-	-	-	0%	-	29
17:30	17:45	24	-	-	1	4%	1	25
17:45	18:00	30	1	-	-	3%	1	31
18:00	18:15	20	-	-	-	0%	-	20
18:15	18:30	33	-	-	1	3%	1	34
18:30	18:45	31	-	-	-	0%	-	31
18:45	19:00	28	1	-	-	3%	1	29
19:00	19:15	20	-	-	1	5%	1	21
19:15	19:30	13	-	-	1	7%	1	14
19:30	19:45	16	-	-	-	0%	-	16
19:45	20:00	6	-	-	-	0%	-	6
Total		426	-	-	6	0%	6	432

Crossing: 5 Movement: L Date: 15/09/2015

Time		Lightweight	Trucks Trailers Buses			Heavy		Total
of	to		2 Axles	6 E	2 E	%	Total	
07:00	07:15	1	-	-	-	0%	-	1
07:15	07:30	-	-	-	-	0%	-	-
07:30	07:45	1	-	-	-	0%	-	1
07:45	08:00	2	-	-	-	0%	-	2
08:00	08:15	2	-	-	-	0%	-	2
08:15	08:30	2	-	-	-	0%	-	2
08:30	08:45	1	-	-	-	0%	-	1
08:45	09:00	2	-	-	-	0%	-	2
09:00	09:15	-	-	-	-	0%	-	-
09:15	09:30	1	-	-	-	0%	-	1
09:30	09:45	1	-	-	-	0%	-	1
09:45	10:00	-	-	-	-	0%	-	-
17:00	17:15	-	-	-	-	0%	-	-
17:15	17:30	-	-	-	-	0%	-	-
17:30	17:45	1	-	-	-	0%	-	1
17:45	18:00	2	-	-	-	0%	-	2
18:00	18:15	-	-	-	-	0%	-	-
18:15	18:30	2	-	-	-	0%	-	2
18:30	18:45	-	-	-	-	0%	-	-
18:45	19:00	1	-	-	-	0%	-	1
19:00	19:15	2	-	-	-	0%	-	2
19:15	19:30	1	-	-	-	0%	-	1
19:30	19:45	1	-	-	-	0%	-	1
19:45	20:00	-	-	-	-	0%	-	-
Total		23	-	-	-	0%	-	23

Crossing: 5 Movement: M Date: 15/09/2015

Time		Lightweight	Trucks Trailers Buses			Heavy		Total
of	to		2 Axles	6 E	2 E	%	Total	
07:00	07:15	5	-	-	-	0%	-	5
07:15	07:30	2	-	-	-	0%	-	2
07:30	07:45	6	-	-	-	0%	-	6
07:45	08:00	6	-	-	-	0%	-	6
08:00	08:15	5	1	-	-	17%	1	6
08:15	08:30	5	-	-	-	0%	-	5
08:30	08:45	3	-	-	-	0%	-	3
08:45	09:00	6	-	-	-	0%	-	6
09:00	09:15	6	1	-	-	14%	1	7
09:15	09:30	2	-	-	-	0%	-	2
09:30	09:45	8	-	-	-	0%	-	8
09:45	10:00	7	-	-	-	0%	-	7

17:00	17:15	6	-	-	-	0%	-	6
17:15	17:30	11	-	-	2	15%	2	13
17:30	17:45	11	-	-	-	0%	-	11
17:45	18:00	7	-	-	-	0%	-	7
18:00	18:15	5	-	-	-	0%	-	5
18:15	18:30	5	-	-	-	0%	-	5
18:30	18:45	11	-	-	-	0%	-	11
18:45	19:00	8	1	-	-	11%	1	9
19:00	19:15	4	-	-	-	0%	-	4
19:15	19:30	8	-	-	-	0%	-	8
19:30	19:45	3	-	-	-	0%	-	3
19:45	20:00	6	-	-	-	0%	-	6
Total		146	-	-	2	0%	2	148

Crossing: 5 Movement: N Date: 15/09/2015

Time		Lightweight	Trucks Trailers Buses			Heavy		Total
of	to		2 Axles	6 E	2 E	%	Total	
07:00	07:15	8	-	-	-	0%	-	8
07:15	07:30	4	-	-	-	0%	-	4
07:30	07:45	9	-	-	-	0%	-	9
07:45	08:00	9	-	-	-	0%	-	9
08:00	08:15	10	-	-	-	0%	-	10
08:15	08:30	12	1	-	-	8%	1	13
08:30	08:45	9	-	-	-	0%	-	9
08:45	09:00	17	-	-	-	0%	-	17
09:00	09:15	7	-	-	-	0%	-	7
09:15	09:30	10	1	-	-	9%	1	11
09:30	09:45	3	2	-	-	40%	2	5
09:45	10:00	10	-	-	-	0%	-	10
17:00	17:15	15	1	-	-	6%	1	16
17:15	17:30	11	1	-	-	8%	1	12
17:30	17:45	9	-	-	-	0%	-	9
17:45	18:00	11	1	-	-	8%	1	12
18:00	18:15	15	-	-	-	0%	-	15
18:15	18:30	8	1	-	-	11%	1	9
18:30	18:45	10	-	-	-	0%	-	10
18:45	19:00	9	-	-	-	0%	-	9
19:00	19:15	14	1	-	-	7%	1	15
19:15	19:30	11	-	1	-	8%	1	12
19:30	19:45	13	1	-	-	7%	1	14
19:45	20:00	12	-	-	-	0%	-	12
Total		246	-	1	-	0%	1	247

Crossing: 5 Movement: O Date: 15/09/2015

Time		Lightweight	Trucks Trailers Buses			Heavy		Total
of	to		2 Axles	6 E	2 E	%	Total	
07:00	07:15	4	-	-	-	0%	-	4
07:15	07:30	2	-	-	-	0%	-	2
07:30	07:45	2	-	-	-	0%	-	2
07:45	08:00	5	1	-	-	17%	1	6
08:00	08:15	3	-	-	-	0%	-	3
08:15	08:30	1	-	-	-	0%	-	1
08:30	08:45	4	-	-	-	0%	-	4
08:45	09:00	2	-	-	-	0%	-	2
09:00	09:15	2	-	-	-	0%	-	2
09:15	09:30	7	-	-	-	0%	-	7
09:30	09:45	6	-	-	-	0%	-	6
09:45	10:00	11	-	-	-	0%	-	11

17:00	17:15	3	-	-	-	0%	-	3
17:15	17:30	5	-	-	-	0%	-	5
17:30	17:45	3	-	-	-	0%	-	3
17:45	18:00	6	-	-	-	0%	-	6
18:00	18:15	7	-	-	-	0%	-	7
18:15	18:30	6	-	-	-	0%	-	6
18:30	18:45	2	-	-	-	0%	-	2
18:45	19:00	4	-	-	-	0%	-	4
19:00	19:15	6	-	-	-	0%	-	6
19:15	19:30	5	-	-	-	0%	-	5
19:30	19:45	4	-	-	-	0%	-	4
19:45	20:00	6	-	-	-	0%	-	6
Total		106	-	-	-	0%	-	106

Crossing: 5 Movement: P Date: 15/09/2015

Time		Lightweight	Trucks Trailers Buses			Heavy		Total
of	to		2 Axles	6 E	2 E	%	Total	
07:00	07:15	2	-	-	-	0%	-	2
07:15	07:30	1	-	-	-	0%	-	1
07:30	07:45	-	-	-	-	0%	-	-
07:45	08:00	-	-	-	-	0%	-	-
08:00	08:15	-	-	-	-	0%	-	-
08:15	08:30	-	-	-	-	0%	-	-
08:30	08:45	1	-	-	-	0%	-	1
08:45	09:00	-	-	-	-	0%	-	-
09:00	09:15	-	-	-	-	0%	-	-
09:15	09:30	-	-	-	-	0%	-	-
09:30	09:45	-	-	-	-	0%	-	-
09:45	10:00	-	-	-	-	0%	-	-
17:00	17:15	3	-	-	-	0%	-	3
17:15	17:30	5	-	-	-	0%	-	5
17:30	17:45	3	-	-	-	0%	-	3
17:45	18:00	6	-	-	-	0%	-	6
18:00	18:15	7	-	-	-	0%	-	7
18:15	18:30	6	-	-	-	0%	-	6
18:30	18:45	2	-	-	-	0%	-	2
18:45	19:00	4	-	-	-	0%	-	4
19:00	19:15	6	-	-	-	0%	-	6
19:15	19:30	5	-	-	-	0%	-	5
19:30	19:45	4	-	-	-	0%	-	4
19:45	20:00	6	-	-	-	0%	-	6
Total		61	-	-	-	0%	-	61

MIX
Papier aus verantwortungsvollen Quellen
Paper from responsible sources
FSC® C105338

Printed by Books on Demand GmbH, Norderstedt / Germany